Getting Started in Your Own Wood
Looking after and caring for a small wood

JULIAN EVANS and WILL ROLLS

With illustrations by
JOHN WHITE and STEPHEN EVANS

Permanent Publications

Published by
Permanent Publications
Hyden House Ltd
The Sustainability Centre
East Meon
Hampshire GU32 1HR
United Kingdom
Tel: 01730 823 311
Email: enquiries@permaculture.co.uk
Web: www.permanentpublications.co.uk

Distributed in North America by
Chelsea Green Publishing Company, PO Box 428, White River Junction, VT 05001
www.chelseagreen.com

© 2006 Julian Evans, 2015 Julian Evans & Will Rolls
Reprinted 2019
The right of Julian Evans and Will Rolls to be identified as the authors of this
work has been asserted by them in accordance with the Copyrights, Designs and
Patents Act 1988

Designed and typeset by Emma Postill

Illustrations by John White and Stephen Evans

Cover design and image by John Adams

Printed in the UK by Bell & Bain, Thornliebank, Glasgow

All paper from FSC certified mixed sources

MIX
Paper from
responsible sources
FSC
www.fsc.org
FSC® C007785

The Forest Stewardship Council (FSC) is a
non-profit international organisation established
to promote the responsible management of the
world's forests. Products carrying the FSC label
are independently certified to assure consumers that they come from forests
that are managed to meet the social, economic and ecological needs of
present and future generations.

British Library Cataloguing-in-Publication Data
A catalogue record for this book is available from the British Library

ISBN 978 1 85623 212 8

Julian's timely update of *Badgers, Beeches and Blisters* (including Will Rolls' chapters on woodfuel) is quite simply a comprehensive introduction to small woodlands and their care for those new to ownership. Written in accessible language, it demystifies management and provides owners with the confidence to make the most of their woodlands whatever their objectives. We'll certainly be recommending it as essential reading.

Phil Tidey
Small Woods Association

This is an introduction for the new woodland owner written by experts in a friendly style, dealing with basic issues from what are the boundaries of the site, what sort of access do you have, to how do you want to manage it. The headings and sub-headings tend to be as questions, with the text giving answers or explanations. At the back is a useful glossary, list of acronyms and abbreviations, notes on common species, organisations to join or consult etc. Throughout, the importance of management is stressed, whatever the new owner's primary objective might be, and the authors have done a good job in showing that getting started in woodland management is not necessarily as daunting or complicated as it may at first appear. The enthusiasm of the authors for their subject comes through, alongside the pleasure that they have got from their woods, and their desire that others may get similar enjoyment. That surely must be part of what the Independent Panel for Forestry in England meant when it stressed the need to see a new woodland culture in which woodlands and wood as a material and fuel are highly valued and sought after.

Keith Kirby
Department of Plant Sciences, Oxford

Getting Started in Your Own Wood has the triple advantage of being co-authored by a professional communicator, a much-respected forester and also a hands on woodland owner where he has learned in "the best of all classrooms – his own wood". This charming book conveys very tellingly the enthusiasm, love and fun involved in woodland ownership. It is written in a manner that is virtually completely non-technical and should be intelligible to everyone interested in woods.

Peter Savill
author of *The Silviculture of Trees used in British Forestry*

About the Authors

Julian Evans

Julian Evans is currently President of the Institute of Chartered Foresters (ICF), Britain's professional body for forestry and arboriculture. He was formerly Professor of Forestry at Imperial College and prior to that Chief Research Officer (S) at the Forestry Commission's Alice Holt research station near Farnham, Surrey. He has written many books and numerous papers on forestry, his most recent of which is *God's Trees – Trees, Forests and Wood in the Bible* (Day One).

For 30 years Julian has owned and managed his own 30 acre woodland in Hampshire and written two books about the fun as well as the frustrations(!) – *A Wood of Our Own* (Permanent Publications) and *What Happened to Our Wood* (Patula Books). He regularly lectures on the subject of caring for a wood. For 10 years he has provided advice to Woodlands.co.uk on silviculture and, in southern England, visits their newly acquired woods to suggest sensible management units when creating parcels for sale. He wrote *Badgers, Beeches and Blisters* (Patula Books) to provide advice to such new owners of small woods that is the forerunner of *Getting Started in Your Own Wood*.

Julian is married to Margaret and has three sons and three granddaughters. In 1997 he was appointed OBE for services to forestry and the Third World.

Will Rolls

Will Rolls is a chartered forester who specialises in the use of wood as fuel. Having worked for a number of years at the Biomass Energy Centre, a small unit within the research agency of the Forestry Commission, he now operates an independent consultancy advising clients on forestry and the woodfuel supply chain. He is author of the bestselling, *The Log Book – Getting the Best from Your Woodburning Stove* (Permanent Publications). You can find out more at www.wrolls.co.uk

Will lives in Yorkshire with his wife and daughter. If you give him a woodland to play in, and a stick to poke things with, he'll be a happy man.

Foreword

Anyone taking on the responsibility of caring for a wood is likely to be somewhat overwhelmed both by the state of his or her ignorance and the amount of advice that can be found online, through networks and in books. Any trepidation is lessened by this volume, which combines excellent references to all that advice, alongside a readable, practical and above all unflappable account of getting on with caring for your woodland. It is a perfect starting point, building the confidence of the new 'parent' who can then find further support elsewhere.

As this second edition is published, more and more people are enjoying the privilege of owning their piece of the UK's natural heritage, in the form of their own woodland. Julian reminds us to learn to interpret its history, to rejoice in the character it reveals year on year and to be bold in enhancing its beauty and value. He brought his now grown up children to his woodlands over the years and has planted in remembrance of dearly loved relatives. But equally he's not afraid to remind us of the value of spending a few pounds on wide gates and controlling the grey squirrel.

I'm delighted to see this valuable volume recast in 2015, to include the opportunities of woodfuel along with threats to the health of our trees. Julian possesses, in life and in writing, a delightful optimism, based thankfully on experience and hard work. A few hours by a fire fuelled by our own logs, reading *Getting Started in Your Own Wood*, will inspire and skill us in equal measure.

Sophie Churchill OBE, PhD, FICFor(Hon)
President-Elect,
The Royal Forestry Society of England,
Wales and Northern Ireland

January 2015

Preface

Julian Evans

Getting Started in Your Own Wood grew out of the very successful *Badgers, Beeches and Blisters* (Patula Books). Indeed not only has every chapter of the earlier book been thoroughly revised and updated, but two new ones have been added thanks to having Will Rolls as a co-author.

Thanks too to Woodlands.co.uk as the sponsor, whose original idea it was for an introductory guide to caring for a wood, and for endorsement by the Royal Forestry Society of England and Wales with their special interest in the well-being of our trees and woods.

What Will and I hope to achieve is a short book that, first and foremost, is enjoyable to read. The book must be factually accurate and address just the sort of questions a new owner or someone newly responsible for a wood might wonder about. More than that the book should become a friend as the adventure of caring for and looking after your own patch brings the fun as well as the vicissitudes of management. So we hope you will find this book full of information, but not overly full of endless lists and instructions; we want to help get you started, not prescribe everything in manual-like detail. You will need more information and so we have provided plenty of suggestions for further reading, organisations to join, and websites to visit.

Will and I both speak from our own experience, but note that the experiences recounted in Chapters 8 and 10 are Will's and that in other chapters are mine.

Enjoy your wood.

Preface

Margaret and Alastair Hanton
Owners of a small wood in East Sussex since 1975

Here at Woodlands.co.uk we are delighted that Julian, with Will's help, has found time to bring up-to-date his popular book *Badgers, Beeches and Blisters*.

Julian has a unique combination of unrivalled silvicultural knowledge and personal hands-on experience. He is president of the Institute of Chartered Foresters and was, for many years, the Forestry Commission's Chief Research Officer (South).

This he combines with 30 years of personal ownership of a small wood, where he has experienced many of the pleasures and tribulations that affect owners. With his background knowledge he was able to overcome most (though not all!) of the obstacles, and to enjoy more deeply the satisfactions. Here he shares with us his expertise and this enthusiasm.

Woodlands.co.uk, which grew out of our passion for woodland ownership, has conservation and enjoyment as a primary focus. We have found that new purchasers come from all sorts of backgrounds but almost universally want to preserve part of this country's rich natural heritage. *Getting Started in Your Own Wood* will help them do just that.

Acknowledgements

We are greatly indebted to Esmond Harris and Alex Argyropulo for reviewing the draft of this book when it was *Badgers, Beeches and Blisters* and making many valuable suggestions. Margaret Hanton's comments were invariably encouraging but also constructive in helping, we hope, to achieve the right tone for our readership.

I would like to thank Will Rolls for coming on board as co-author and all his careful and detailed review updating the text as well as sharing his expertise as one of Britain's firewood and wood fuel experts.

We continue to be very grateful to John White for his wonderful illustrations although the ones of the tractor (page 29) and of the logs (page 83) are by my son, Stephen.

Contents

1

Owning or Caring for a Wood

I can remember lying awake the night I bought my own wood wondering, and worrying, if it would be OK.[1] It was wholly irrational: after all the wood had survived perfectly well without me. Now I was its owner and carer and, well, it felt a little like when a new born baby has its first night at home and you tiptoe to the cot to see if the mite's still breathing. The excitement of buying a wood, of owning your own patch, or simply becoming responsible for one, is something to savour, and in a sense not worry about at all but enjoy. This short book is to help you do just that.

The early days...

The early days of getting to know a new wood are full of surprises. The changing scenes as sunshine turns to rain, your first storm with trees swaying and rocking, or your wood draped white and sparkling in a brilliant mid-winter's morning after a night of snow. The seasons themselves are counted off by flowers and herbs – primroses, bluebells, herb bennet, pimpernel in the rides, mushrooms and toadstools of autumn – by the rise and fall of birdsong and the deep silence of August, by your trees as their buds burst and new leaves build crown and canopy, and then age, turn russet, golden, or brown and finally fall to the forest floor only to begin again in the spring. I could go on for the seasons never fail to surprise and delight as on each visit your wood looks different or reveals a new facet. What stocks and shares can ever match that?

But surprises do not only reveal themselves when you are getting to know your wood. Even after 30 years of changing seasons we find new things. Several years ago it was that lovely

---

1   In all chapters except the two Will wrote (8 and 10) the 'I' refers to the first author, Julian.

wayside plant of central southern England, Solomon's seal, and then it was a clump of helleborine under some beech trees right beside our entrance. For several days in late May we had worked hard and raised blisters setting posts, attaching rails, and replacing the main gates. During a coffee break there they were, two patches of this close relative of the orchid – wonderful, and I'd only been going in and out of the entrance for nigh on two decades! A surprise of a different sort is learning that yellow archangel isn't simply a golden version of the similar looking white dead nettle, but an indicator of ancient woodland habitat. And discovering that muntjac devour bluebells and target the best patches of cowslips.

You will realise I am a bit of a fan of wildflowers. We also have buzzard nesting though they haven't always, badgers are in the vicinity but not yet active in the wood, a marsh tit was spotted by friends and the red kite is now commonplace, orange-tip butterflies herald spring warmth, and deer, both roe as well as the wretched muntjac, are increasing. At least they were, my wood is now deer fenced, but I'll come back to that later. Woods are never static; they are forever variations on a theme.

Being creative...

The pleasure of wildlife, of being in the country, need not exclude more down to earth interests of cutting firewood, gathering pea and bean sticks or growing useful timber. Unlike a farmer's field devoted exclusively to wheat or oilseed rape, your wood can meet many needs simultaneously, and we will return to this theme as a helpful way of thinking about woodland management. The pleasure this brings is that while you can enjoy the changing scenes in your wood from the weather, the butterflies, birds and flowers in their season, you can also create change. You can plant, and fell, you can pollard and coppice, you can deliberately leave piles of dead wood for beetles or open up glades for picnics and even dig ponds. The scale of such interventions needn't be timid, indeed they usually shouldn't be. It's a question of confidence. A wood is so permanent, trees are so tall and timeless, dare you interfere? Yet interference, informed interference if you like, is just what many woodlands cry out for to help regenerate them, to help maintain the very diversity you are afraid of losing, and even to earn some money. Over the years our patch has more than 'washed its face' financially, but for many that is not an important consideration. Walking the dog, camping with the kids, pretending to be Ray Mears, fashioning your own rustic furniture, or having your own resource for woodturning are all reasons enough.

The possibilities woodland affords are almost endless. There's getting to know your neighbours – always good to be on friendly terms – finding out who locally has special knowledge to help, who in the county can provide advice, is there a farmer who can mow or swipe the rides in summer in exchange for some Christmas trees in winter and so on. And have you a skill to offer or now, with your wood, a place to invite friends and family to?

The handsome nuthatch forages in cracks and crevices in the bark mostly working down the trunk, the brown and better camouflaged treecreeper does this only going upwards.

Woodland does you good

As well as the value of exercise, there is now good evidence that a green environment helps recovery from illness and helps us relax and ease stress. Indeed, Woodlands.co.uk published a report in 2013 entitled, 'Happiness grows on trees'. Based on a survey of woodland owners and review of published work they subtitled the report, 'How woodlands boost our wellbeing'.

How much time does it take?

I often give talks about our wood and the most frequently asked question, by quite a long way, is how much time do you have to spend there or how much time does it take to look after properly? There is no fixed answer. Unlike a dairy farmer's daily necessity of milking cows, a woodland can be left, and left, and left – though presumably you won't be enjoying it as much as you might! Apart from checking the entrance, which someone else can always do for you anyway, or occasional visits for pest control if you have a particular problem such as grey squirrels, there is nothing in woodland management that requires you to visit this week or this month. You may choose to, but will not have to. You can visit your wood when you want to, not because you must.

In answer to the question for myself, over the years I have probably averaged half a day per fortnight in my 30ac wood. This is very frequent because, as I suspect you already appreciate, my wood is my hobby and I derive hours of pleasure from caring for this gift God has entrusted to us. I can honestly say that I have never found it a chore to visit my wood: I hope you find the enjoyment of yours as rewarding.

2

First Steps

To care for a wood, indeed to enjoy it to the full, you need to get to know its ins and outs, literally as well as figuratively! Like a new house it's more than knowing the number of bedrooms or how modern the kitchen is: just as you would look into hidden corners, poke into places not usually probed and actually investigate what's behind the garden shed, it is good to take time to become familiar with a wood. Of course, it is possible to ask a forestry consultant to prepare a report, like a surveyor evaluating a house, and their professional descriptions and judgements are particularly valuable for matters of commerce, law or safety, but that's not quite what I mean. It is hugely rewarding to gather firsthand knowledge about a new wood: here we look at the basics. Even so, however long you have it you will always be finding new things.

Getting to and from your wood

One of the essentials you need to check at the outset is the rights of access from a public highway – and we will come back to that – but as well as this there is the simple question of how far you live from your wood? It is a more interesting matter than you might think.

If your wood is next to your home or only a mile or two distant, you can pop over at whim, walk the dog, or enjoy a picnic as fancy takes you. And it's easy to check the entrance for rubbish. The converse is that you might find yourself spending many more hours there than intended and your partner or family become whatever is the silvicultural[1] equivalent of a 'golfing widow or widower'. Also, a wood that is nearby can be readily

---

1 'Silviculture' is the forestry equivalent of agriculture: 'agri' is from the Latin for field, 'silvi' for woodland, so silviculture is everything to do with the care, husbandry and growing of woodlands.

inspected after a storm or when heavy snow has fallen, which can be important if it enjoys roadside frontage or there is a right of way through it and some clearing up is needed.

When a wood is a long way away, say 30 miles or more, then the time taken to get there becomes significant. You can't so easily make a quick visit, it is more of a planned outing or even a day trip. Now, this is not all bad since the visit takes you away from home and away from the familiar: it is more of an expedition for the kids, it is more like going on holiday. So there are pros and cons and this will, in part at least, inform the ways you intend to enjoy your wood.

There is, too, the extreme where woodland or forest has been purchased purely as an investment. It doesn't matter where it is, you can still camp there or take a caravan, assuming acres of Sitka spruce are as congenial to you as, with luck, your bank balance.

For many years my own wood was about 15 miles from my home in Alton, but 10 years ago I moved house and it is now only eight miles away. I found that even 15 miles meant that you can get there in under half-an-hour, do a decent morning's or afternoon's work, and still have the rest of the day for other things. It was far enough away when going for a picnic to make you feel it was a proper outing without the boys getting tetchy in the car, but not so far that the visit had to be planned in advance. It was a nice compromise and, for me, a bonus was that my work often took me past the wood *en route* to Oxford or the West Country so I could quickly check it then as well.

So, how good is the access?

The distance to your wood is a matter of convenience; getting in and out of it from a public highway is a necessity. There's obviously little point having a wood in the middle of a field if the farmer only allows access, other than perhaps on foot, in the couple weeks in September between his harvesting one crop and sowing the next. I am not exaggerating: this has happened to some unwitting owners. To be able to manage and enjoy your wood you need good access, but what does that mean?

• First, good access means you have comprehensive rights.

You should have the right to use the access road or track unhindered and uninterrupted whenever you need to. It also means you can use any kind of vehicle you might conceivably want to, such as laden timber lorries of up to 30 or 40 tons. Even if you yourself don't intend to work your wood commercially, future owners might, so securing this right of access is important.

- Secondly, good access means a track that is wide enough and without sharp bends that are tricky to negotiate. Its formation should be sufficiently load-bearing to support the biggest vehicles ever likely to use it.

- Thirdly, good access means a wide and generous entrance on to the public highway, either where your wood fronts on to it directly or where the track you have rights over meets it. Articulated lorries, or a car and caravan, cannot turn in easily. From a lane the gates need to be well set back, gate posts at least 6m apart and the whole entrance bell-mouth opening out to perhaps 30m across – see the illustration at end of the chapter (page 16).

My own entrance before it was renewed to the standard outlined here.

I know all this is the counsel of perfection, but hopefully it will help you think what you have got at the moment and what improvements may be needed one day.

Lastly, good access means good internal access within your wood: a desirable rather than essential feature. Are there tracks and rides – these words are often used interchangeably – that allow access to all parts? Is there a turning area and loading bay beside or at the end of the main access track from the public highway where logs can be stacked or a caravan or visitors' cars parked safely? In general, the wider the track the better. Wide tracks dry out more quickly, offer more 'edge effect' for wildlife, and are great places for kids to play.

As you get to know your wood, note these features about access, and why not also think about a new footpath, perhaps with a little mystery, that winds its way past an old wizened tree, brings you to a view, or takes you to a secret glade? I have one in my own wood, not so much by design, but the route I usually take with first time visitors or when we have an Open Day.

Public access and anyone else with rights

The deeds should show whether there are public rights of way across your land, as should the appropriate Ordnance Survey map. The definitive statement is held at the County Record Office. I assume you have already bought the OS 1:25 000 scale (Explorer series) that covers your wood, but don't forget the larger scale 1:10 000 maps if available and even the 1:1250 as a basis for a woodland plan or map. Indeed, any applications to the Forestry Commission (FC) will need to be made using the OS 'Mastermap' for the site (1:1250). Maps of your woodland are available from your local FC office to support the application process. (As an aside, you can get your very own Ordnance Survey plan from Stanfords for a small fee with grid lines and contours included for a little extra.) The County Record Office will let you have copies of earlier maps – the fabulously detailed 1870 series is a must if you have an interest in the wood's history.

Returning to the question of public access, whether footpath, bridleway, or byway such as a green lane, there are attendant duties on the landowner that should be checked.

However, you may discover that others have access rights either specified in the deeds or when strangers start exercising them, as happened to us! These are mostly of three kinds:

- Rights to sporting, such as rearing and shooting pheasants and other game birds, taking of deer, and even rough shooting – rabbits, pigeons, grey squirrels etc.;
- Where a utility crosses the land, such as electricity or telephone, and a 'wayleave' has been granted, known legally as an 'easement', which may have a few restrictions;
- Specific provision for access – in my own wood Network Rail have the right at all times to visit their electricity transformer and gain access to the railway track at the bottom of the wood. We share the same lock on the gate and from time to time they contribute to upkeep of the main track. For my part I must ensure that this track is never obstructed.

Many woods have none of these additional rights and some rights, such as sporting, may only be for a period and you, as the new owner or manager, can usually change the terms or stop their exercise altogether.

Locks, gates and names

A gated entrance that looks tidy and well kept conveys a sense of pride, care and regular usage, all of which will help to deter fly-tipping and other rural crime. The type of gate is unimportant, but the common metal or wooden 'five-bar' gates are very serviceable and rarely appear out of place in the countryside. To deter theft, particularly of new gates, use one hinge upside down so that the gate can't simply be lifted off. A substantial lock and chain add to security and give a business-like impression, but do make sure that all who have a right of access have a key for your lock! It's not uncommon for several owners to share access and a locked gate. Do try and get everyone zealous in keeping it locked.

If your wood has a name you can attach a board or plate to the gate, or erect a free-standing sign. Of course if you use a shared entrance this might not be possible. And there's no reason why you can't name parts of the wood inside. We now have a 'Taid's Wood' and a 'Nain's Copse' in our 30 acres, named incidentally, after my parents. 'Taid' is Welsh for grandfather, 'nain' for grandmother: the story behind these namings are in the two books I've written *A Wood of Our Own* and *What Happened to Our Wood*.

Personally I dislike signs like 'Private', 'Keep out' and the asinine 'You are being watched'. We no longer see so much the more threatening 'Trespassers will be prosecuted' apart, of course, from the pages of *Winnie-the-Pooh* and Piglet's pride in is his grandfather 'Trespassers Will…'!

Sheds and things

You don't need to have a shed in the wood since saws, axes, spades, garden chairs and other paraphernalia are all easily carried in the boot of a car. If you are there daily then a shed is convenient, but it is debatable whether it is worth locking since someone will doubtless find it and doubtless want to break in, so don't keep anything of value in it. Arguably a trailer for your car is a better investment if you are planning to do a lot of work in your wood.

That said, if you plan to spend a good deal of time at the wood, some furniture is useful. Old tables and chairs are handy, but can quickly turn a pleasant glade into a slum. Despite what I have said, it is neater to keep them in a small shed along with kettles and dishes, spare wellies, tarpaulins etc. Fasten the door, but don't waste money on an expensive lock, it will simply attract not deter interest. Thieves are looking for power tools, not your old furniture!

Remember that if you are planning to use herbicides or poisons, say for weed or rabbit control, then a secure, lockable safe may be required by law and is probably best kept at home in the garage.

Boundaries

Walking the entire perimeter of a new wood is one of the first things to do. It's fun. Apart from discovering remote corners, defiantly inaccessible bits, patches of nettles, and doubtless some rubbish, it will alert you to several things: what the state of the fencing is; where animals (and people) may be gaining unwanted access; any trees next to a highway that may look unsafe; whether a neighbour is using your land; and, of course, whether the line on the deeds matches where your boundary appears to be! Wear tough clothes and be prepared to fight through undergrowth in order to follow the exact course. There's no need to do it all in one go.

Maintenance of fencing will depend on need and whether the wood has a covenant requiring you to do this. This is more common than you might think since in the 1930s and 1940s, when the Forestry Commission was acquiring much land and some already established woods, as a gesture of good will it agreed to fence out a neighbouring farmer's livestock and so take on the job of fencing. It's back-to-front, after all it is the farmer's sheep and cows that move about and not trees, but quite often today part of a wood's boundary may still having a fencing covenant to be met by the owner. This may have lapsed or a change of neighbour extinguished it, but it's worth checking.

Fencing is an interesting countryside skill. It is one that can, in time, be learned by almost anyone, but is also one where there are many contractors who will come and do a thoroughly professional job. Your main decision is whether the fence is purely for demarcation, where an attractive post and rail or simple strand wire will do, or one that is to keep out livestock, deer or rabbits in which case wire netting of suitable mesh size will be required – we go into more detail in Chapter 6. Doing your own fencing from coppice products can be very rewarding but make sure stakes

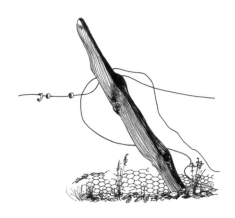

Old oak post with remnants of rabbit netting – only maintain a fence while the wood is at risk; remove if it becomes unsightly.

are durable, either treated softwood or, better, sweet chestnut, oak, or even Lawson cypress from an unwanted garden hedge.

What is the woodland like?

This chapter is almost finished, and we haven't got round to talking about the trees themselves and discovering what they and the wood as a whole are like! I expect your vendor's particulars told you the basics, and perhaps that is what most interested you anyway. We will look at five questions.

'All sorts' or one sort?

I'm not referring to Bassett's lovely liquorice sweets, but is your woodland diverse or fairly uniform? Do you have young and old trees, dark and dense stands and light sunny gaps, wet places and dry slopes, and many kinds of tree species, or is there little variety with perhaps one or two stands of, say, Scots pine all of the same age? We British have been great planters of trees over the centuries and one result is that we have created woods and forests that tend to be uniform – stands that are even-aged and of only one or two species.

The importance of this question, what we call the structure of a woodland, is the potential it offers as an amenity or for wildlife – generally the more diverse the better – or for commercial use when less diversity usually has the edge economically. Of course, your management over time can coax a wood towards either more or less variety.

Is it an 'ancient' woodland?

By ancient I mean: Has the land always been wooded? It is considered to be so if early maps or other records show it to have been continually woodland from before 1600. If it was then it almost certainly always has been.

The significance of the question, quite apart from the piece of living history you might own or be responsible for, is that ancient woodlands are limited in extent and are usually the richest in wildlife. Indeed, some woodland flowers such as wood anemone, yellow archangel and oxlip are confined to them and help indicate this special status. For this reason your management options may be curtailed somewhat. For example, as a condition of tree felling you will usually be allowed only to plant or regenerate native tree species.

We return to the question of 'ancient' in later chapters

because it is important, it affects what you can do, and as relics of our past woodland type they are to be treasured.

How has the woodland been managed in the past?

Is your wood a plantation, a coppice, coppice with standards, or neglected wood pasture, perhaps with pollards etc.? It is not always easy to tell, especially when a wood has been neglected many years or even many decades. Knowing its silvicultural history will help you since most coppices can be restored to working order in this way (see page 74) while woodlands that have been planted probably need to be regenerated in the same way.

At the end of the chapter (pages 17–18), the main woodland types and their tell-tale signs are illustrated so that you can be your own detective.

What are the main species?

I expect the vendor's particulars will have answered this, but any good tree identification book will help with the more common species. One reminder for the uninitiated is that all conifers are known as 'softwoods' in the timber trade and all broadleaved tree species as 'hardwoods'. The terms have nothing to do with how hard or dense the actual timber is: yew is a conifer and is a 'softwood'; birch, poplar, and even the tropical 'balsa', are broadleaved species and are 'hardwoods'!

What commercial potential might it have?

The best way to answer this question is to invite a professional forester to visit your wood for an hour or two. He will look at the species, the ages and size of trees, how many hectares are ready for thinning, felling or coppicing and so on. For example, one or two high quality trunks of broadleaves (oak, ash, wild cherry etc.) can be worth enough for a merchant to buy as individual trees or logs. In this instance 'high quality' is a straight, defect free trunk of large diameter (50cm+) able to produce a log of 3–6m in length. More usually however, in any wood including plantations of pine, spruce or beech, being able to make up one load (15-20 tons) is the very minimum that might attract a purchaser – we return to this topic in Chapter 7.

In coppiced woodlands it should be more obvious whether it has recently been worked and the vendors, your neighbours or local forestry people can probably tell you. This is revealed in the wood itself by a patch (0.5-2.5ac) of young growth next to older coppice – again, refer to the illustration at the end of the chapter.

Remember though, as we noted earlier, as well as the trees themselves, commercial potential will be dictated by the question of accessibility.

Wildlife surveys, archaeology, and censuses

For me one of the pleasures of owning woodland is finding out about wildlife, archaeological features, and how the wood has changed over time.

For recent history the 19th Century tithe maps, old records and early aerial photographs are invaluable. A trip to the local County Record Office is a must. However, I wonder how many people know that every piece of woodland in Britain over 5ac was inspected for the Forestry Commission's 1947 census of forests and woodlands. You can inspect the original report on your wood at the National Records Office at Kew. Mine was visited on 19th June 1947 and was described as 'devastated' (thanks to wartime fellings) and, rather quaintly, suitable for 'economic management'.

Surveys of wildlife are on-going and in one sense are never complete and always bring surprises. You can simply enjoy spotting wildflowers, butterflies and birds and build your own nature notes as season follows season. All of us probably know a 'twitcher' or two who will identify this bird song and that type of nest, even if not up to the naturalist, Chris Packham's standards.

"That's never a golden oriole!"

There is much history to be discerned from archaeological features, such as banks and ditches, and of the woodland itself from the appearance of the trees. Numerous local societies will be more than keen to visit your patch and provide advice and information.

My records are quite informal, just a few jottings in a notebook. When I was writing this chapter for *Badgers, Beeches and Blisters* I reported that, "Only yesterday I found the first ever ragged robin flowering among nettles in damp ground right beside the main track. My wife, Margaret, had sown wild seed and planted a few tiny seedlings the previous July, and if I hadn't been doing ride-side maintenance it would not have been seen." But enough of me: do record what you find whenever you can, note any hazards and dangers that need attending to, and do rope in friends and relatives and their knowledge to enrich yours.

Peacock butterfly with its favourite food, the stinging nettle.

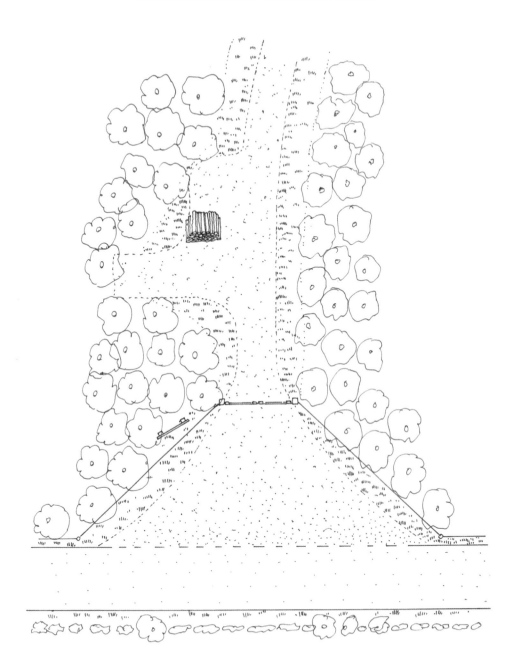

A well proportioned woodland entrance.
Gates should be set at least 6m back from a highway, but check local
planning guidelines first, and the frontage about 30m across
to allow large vehicles to turn. The gates themselves
should provide an opening of at least 5m.

The main types of woodland:

(i) A young plantation. Trees evenly spaced in rows and all of much the same size.

(ii) Weeding and cleaning a plantation to stop it being overgrown.

(iii) Thinning a plantation to give space to the best trees and yield some produce.

(iv) Clearfelling underway (right) and replanting (left)

(v) Coppice with standards

*(vi) Wood pasture – these are the kinds of trees found but are farther apart
– more scattered – than John White's delightful frieze suggests.*

*(vii) Pollards you might see. From left to right: managed,
neglected, ancient, and riverside willow.*

*(viii) Continuous cover forestry – new name for an old system
which ensures that woodland is never completely cleared.*

3

Enjoying Your Wood

Thirty years ago I wrote a book for the Forestry Commission on how to grow and look after broadleaved woodlands.[1] In the book the question of what is the principal object of management was emphasised as the first thing to decide when considering what to do with a wood. This is still the case today. And the title of this chapter is just a way of saying it! Taking time to reflect on objectives or reasons for having a wood is constructive, even liberating.

We will look at the main ways any woodland can be enjoyed, highlight the issues, and suggest steps to take to get you started. But, as was emphasised in Chapter 1, different aims are rarely mutually exclusive, a wood can satisfy several at once, and no two woods are the same. What we tease out here are the main hopes you have to help decide what should take priority in the way you care for and look after your wood.

Why we like woodlands

Three years ago Woodlands.co.uk commissioned a survey, 'Why people buy small woodlands and how they manage them'. Of the 149 owners most were active in management, visited their wood frequently, placed conservation and wildlife as top priority, and over three-quarters intended to pass their wood to future generations of their family. While this survey focused on only one sector of woodland ownership and care, it betrays the sorts of reason why we like woodlands.

There is another reason, perhaps less obvious, that is linked to personal well-being. There is now solid scientific evidence to show that a green environment does us good. Recovery times in hospital are improved when the patient looks out on 'greenery'

---

1 *Silviculture of Broadleaved Woodland*, Forestry Commission Bulletin 62, HMSO/TSO 1984.

compared with a built environment. Conversely the loss of millions of ash trees from the streets of many American cities, because of emerald ash borer infestation, led to a significant and measurable rise in human mortality.[2]

To get away from it all

For many enjoying a wood for walks, picnics, and simply to have fun is the main reason for having somewhere in the countryside which is their very own. Getting away from it all and being able to relax are greatly valued in today's pressured times.

An attractive woodland for this purpose, what professionals call 'amenity', is one that has variety in tree sizes and ages, plenty of tracks and paths, and lots of open space. If you have a stream running through the wood so much the better, though do be careful with any watercourse and ponds if young children are about.

If peace and quiet are particularly sought after, then being some distance from a motorway, major arterial road, airport or railway line may be important. From experience if you can see from your wood where a major road is, even if two miles away, you are likely to hear a distant rumble of cars and lorries if the wind is coming from that direction. Of course, you will always have farm machinery and other countryside sounds: in our wood we have microlights buzzing overhead from a nearby airfield, not to mention the railway line at the bottom, but neither really intrude into our enjoyment – indeed, my boys when young loved to watch the trains.

It is also good if such woodlands are not too exposed, open and windy, though obviously you can choose when to picnic; after all you don't need to go on a stormy day or when a cold easterly wind is blowing!

Other points to think about – and we will return to this in Chapter 4 – are:

- Elderly people like to feel safe, so for them in particular paths need to be even under foot so they won't trip or fall. They also need somewhere to sit;

2 Donovan et al. (2013) The relationship between trees and human health. *American Journal of Preventive Medicine, 2013*; 44(2): 139-145.

- Most kids, will climb trees, make camps, gather sticks, and get stung as they muck about, so a first aid kit is essential and a mobile phone desirable assuming you get reception in your wood;
- Fires should be restricted to one or two safe locations well away from the base of a tree;
- Glades make great picnic sites and double up well for wildlife;
- Vistas need preserving and maintaining – a wonderful view from the wood may not last unless the opening between the trees or shrubs is kept cut back;
- Tracks will need to be cut or mown once or twice each year;
- A small shed for gear, and if there's a sudden downpour, but don't store power tools and anything of value.

'Rest awhile'

If you plan to camp in the wood, then how easily can you get your gear to the chosen site may be important – is the track only adequate for fair weather use for the family car? There is the question of disposing of waste: taking it home usually being the best option.

What to concentrate on

If enjoying your wood as an amenity for recreation is your main aim, then the key things to concentrate on are providing good, safe access (tracks etc.), knowing the dangers and hazards, and making sure you have some means of communication in case of accident. Of course you will want, over time, to manage it in ways that keep or add to the wood's diversity and interest for old and young alike. We will look at the particular question of insurance and today's need for a risk assessment in Chapter 4.

Something of beauty

More often than one might think, a wood is acquired to preserve it as part of the countryside. A farmer might buy a wood adjoining one of his fields (or plant one on his land) to improve or retain a particular view. Similarly, someone who has moved to the country may value the nearby spinney or copse simply for its place in the landscape or because it screens an eyesore, but not be greatly fussed about visiting it for walks or for having a picnic.

When you buy woodland to preserve the beauty of the countryside it may also be a good investment. A house or cottage in the country that is attractive with pleasing views and free of anything unsightly or smelly(!) nearby, will be worth far more than one overlooking (say) a transport yard. All that might be needed to make the difference is retention of a patch of woodland, even as small as a quarter of an acre. Unfortunately (or fortunately) buying a woodland with the hope of building a house in it is virtually impossible planning-wise,[3] though buying a cottage and later purchasing some adjoining land such as a wood is certainly feasible.

Maintaining the fabric of our lovely countryside is also the aim of bodies such as the National Trust, the Woodland Trust and the Council for the Protection of Rural England (CPRE). The Forestry Commission (and Natural Resource Wales) take great pains to blend their operations into the local landscape with carefully crafted forest design plans and expects others to do so too. The benchmark is set by the UK Forestry Standard that

---

3   A remarkable but rare exception was Ben Law's experience at Prickly Nut Wood featured in *The Woodland Way*; Ben Law, Permanent Publications, Hampshire, 2001.

helps us demonstrate that our woods are properly and sustainably managed. Landscape architecture has come a long way in recent years and much sound advice is available.

Enhancing the attractiveness of woodland

For the smaller wood, the following points should be born in mind where conserving beauty and being a feature in the landscape take first priority. Build on natural features and the form of the land such as rivers, streams, boggy patches and rocky places. Work with nature seeking to enhance your woodland's natural assets by not planting too close or by deliberately felling trees nearby to emphasise the feature. In general native species are preferred over exotics, but don't be slavish is pursuing this. A few pines, larches or Douglas firs in an otherwise broadleaved woodland usually add interest and certainly enhance the variety of habitats; what ecologists call 'niches'. Similarly, don't be over-zealous in removing sycamore; it looks good in the landscape and adds to biodiversity. And don't forget we have three native conifers: Scots pine, juniper, and yew. We return to these topics in Chapter 9.

Think about the land form, and then manage the wood to be sympathetic to slopes, shapes and appearance of the local topography and texture of the landscape. A square block of dark conifers stuck on the side of any hill will generally offend and intrude: the same size of wood with two or three small patches of broadleaves in with the conifers, some open space, and boundaries that are not too geometric, usually will not. This is a generalisation, but I expect you get the point.

The same goes for the shape and size of openings made when felling trees. All woods need some tree felling to maintain them – long-term neglect is rarely the best option, but fell areas sensitively or explore continuous cover forestry options (page 18). As always the aim is simply to look part of, and not intrude in, the landscape.

John White has tried to illustrate for me the above points in the accompanying sketches on the following page.

The internal landscape of a wood is important too. Rides and tracks that curve, have scalloped edges i.e. have occasional glades to one side, and have vistas. They all add interest and are generally good for wildlife.

Attractive landscape where trees and woodlands enhance the natural form

Unsympathetic planting and felling that intrude into the countryside
with harsh edges and unnatural, geometric shapes.

Paintballing and related activities

I am not sure this topic should follow 'beauty' quite so closely(!), but it did so in my thinking so let's run with the juxtaposition. Clearly buying a woodland for activities like paintball games, motocross rallies, or off-road 4x4 training (as Landrover have), bring different management priorities. Key among these will be access rights, insurance cover for groups, prior permission from the local police for the traffic levels expected, in many cases planning permission from local authorities, and avoiding nuisance to your neighbours. Obviously one wouldn't buy a wood of special conservation merit for these inherently more destructive uses, but even so, a wildlife survey and full, transparent discussions about intentions will mollify neighbours and complaints from locals about the inevitable noise as well as allow special habitats to be avoided. It all makes for good public relations and is good stewardship.

Many owners of or those caring for a small wood will have no interest in these sorts of activities, but this does make the point that woodlands of all types are great places for fun and games.

Game and hunting

For many game shooting and hunting will not figure at all in a woodland's life or plans, but it's as well to be aware that woods, including small ones, are valued by these very common countryside sports. Many woods on estates and farms are managed for game such as pheasants and the steadily rising numbers of deer bring a need to stalk and hunt or simply cull. For game birds a wood provides somewhere to rear them, cover and warmth for the birds in winter, whilst areas of young growth and shrubs create flushing cover to help the birds rise to the guns. Keepers may put down supplementary feed in a wood to attract birds in from the surrounding fields and improve numbers that beaters subsequently encourage into flight.

The management of woodlands for game is straightforward and the Game Conservancy at Fordingbridge is undoubtedly the best point of contact for those newly interested.

Rough shooting for rabbits, wood pigeons, grey squirrels and other wildlife officially designated as 'vermin' is hardly likely to be a major management aim, though a local shoot might acquire a wood with this purpose in mind. As with all shooting, compliance with gun licensing laws and exercising of very great care when in your wood are essential.

Deer stalking and culling is a matter for experts. Make local enquiries about who might be suitable, agree terms with them, and from time to time you may enjoy a haunch of venison from your own patch. Have no qualms about culling: deer numbers are at an all time high and are doing serious damage not only to woodlands but many national nature reserves owing to excessive browsing and grazing.

Your own firewood supply

Any woodland, except a new planting, can provide firewood. In these times of energy concerns and the need to consume less coal, oil, and gas, using firewood for heating, such as with an efficient wood-burning stove, (and cooking if you have a suitable Aga or Rayburn) makes sense. Firewood is best obtained from thinning out the poorest trees, from the debris left over after any tree felling, and directly from coppicing operations. The key point is that the shape and, to some extent, the size of the tree is largely irrelevant: everything can be used. This distinguishes firewood from all other timber products where straight, defect free trunks are usually essential.

Will Rolls explains this increasingly important topic in Chapter 8.

A haven for wildlife

Chapter 9 is devoted to this topic, but if protecting or enhancing wildlife in a wood is your main interest, the key issue, beyond finding out what is already present as we noted in Chapter 2, is understanding woodland ecology. Plants and animals all interact – the web of life – and your interventions should be informed by what you know. For example the fabulous fritillary butterflies, which can be so helped by timely coppicing, need a continuous supply of new, open spaces every few years since the plants they feed on mainly inhabit glades and warm sunny patches. So maintaining a wood for its fritillaries should really be seen as a commitment to coppicing or felling openings perhaps every other year. I've over-simplified the matter, but it's fun becoming a real wildlife 'geek' and reversing the downward trend so apparent with so much of Britain's flora and fauna.

Silver-washed fritillaries can be seen in open woodland in high summer, their larvae feed on violets in the spring.

Of course, you won't always succeed first time. Margaret and Alistair, who are supporting this book, told me of a delightful occasion when they set up a hide in their own wood one night to watch badgers. Let me relate it in their own words:

"We set up a hide one night, with a cotton sheet, which we put in place during the afternoon. At nightfall we crept along and stationed ourselves on the chairs we had made ready, and sat stock-still peering through holes in the sheet for an hour or more. Not so much as a black snout appeared from the sett. At last, Alastair said, 'I'm feeling terribly sleepy' but the only answer was a muffled snore from his wife.

The next weekend some friends camped in our clearing. They reported that while they were frying up their supper they received a most surprising visit... from one of our black and white friends... not too shy to enjoy the proffered titbits of bacon!"

Some people have all the luck!

Christmas trees

I do hope you haven't bought a wood just to grow Christmas trees, but I do hope you remember to have a patch of them somewhere in your wood! Prices of real ones today are exorbitant, while plastic ones are made from non-renewable oil. So grow your own.

Plants of the commonly used species, Norway spruce, (never use Sitka spruce)[4] can be bought from a forest nursery for 20-50p each and should make usable size in five years. If you can't wait and have a patch of conifers already in your wood, the tops of slower growing trees will often be suitable – fast growing ones will look very leggy as Christmas trees with widely spaced whorls of branches. Don't be afraid to cut a tree just for its 4-8ft top: what you save in money terms will be far, far more than you are ever likely to make from selling the timber!

Christmas trees that don't drop their needles are becoming much more popular and the main species are Nordmann's (Caucasian) fir, noble fir and Fraser fir. A bushy, slow grown specimen of our own Scots pine always makes a good Christmas tree.

Although we will look at tree planting in Chapter 6, if you want to get started with a patch of Christmas trees the key things to note are:

- Use land that may be unsuitable for other purposes, say because of a powerline overhead
- Keep the patch hidden from the entrance to avoid theft
- Fence the patch against rabbits and deer
- Space trees about 1m apart, on the triangle if you can to make more even contact with neighbouring ones
- Keep weeds under control
- Just help yourself to the biggest as they reach the right size, then restock when two or more adjacent ones have gone and there is a gap of 3m across

Do be aware, however, that Christmas trees are specifically

---

4 Will Rolls says this is the classic forester's comment! He doesn't like it in his home either – it's much too spiky – but it is on sale at Christmas often labelled as 'blue spruce'. I would still avoid it.

excluded from most Forestry Commission (and Natural Resources Wales) grant schemes, and you may find yourself in trouble if you want to plant large areas of them in an existing woodland.

Even timber production!

Here is my son's sketch of the time we extracted over 600 tons of pine from our 30ac wood. By 'we' I mean the contractor we sold them to! The pine trees themselves were 35-years-old and the contractor felled the trees, extracted the logs, and hauled them to market. What we got from the sale almost equalled the price we had paid for the wood seven years before.

My son Stephen's sketch of the tractor with its grapple stacking logs at our wood's entrance when we sold the pines in 1992.

Since then we have thinned out the beech trees three times but we've got nothing like so much either in total quantity of timber harvested or in price per ton until the last thinning. Mid-term (mid-rotation) beech has few markets and initially ours went to a mill at Sudbrook in Gloucestershire for pulping, though this mill has since closed. Fortunately demand for firewood has more than substituted for this lost market and prices have risen markedly.

If timber production is one of your aims, then it's best to get an expert to cast a professional eye over your woodland. An hour or two of their time will soon tell you whether you have some commercial potential, when returns can be expected, and tell you how to go about the business of marketing the wood. I have stressed before that management for this purpose need not exclude other objectives.

In Chapter 7 we look at several topics relating to timber production to give you a feel of the possibilities you might have, though don't be optimistic: prices have been depressed for some years now, and while some markets are growing e.g. firewood, many others are still suffering. There we look at: (1) The rewards of using you own timber whenever you can; (2) How to sell trees to contractors; (3) Advertising trees for sale in magazines, on-line, and via websites; (4) How to assess quality and quantity of timber; and (5) Typical prices for timber.

How else might you enjoy your wood?

The website www.woodlands.co.uk has a splendid list of things to enjoy in your wood. Obviously not all woods are suitable for everything listed, but here is a flavour to round off this chapter nicely (I've added a few of my own):

Exploring, barbeques, picnics and parties, map-making, creating tracks and paths, constructing benches, building secret dens, hanging a swing, burying a time capsule, orienteering, scout and guide wide games, charcoal making, bodging and turning green wood, photography, sketching, nature trails – including a hide to watch birds or badgers, developing a forest school, putting up bird and bat boxes, gathering wild foods such as blackberries and some fungi, bee-keeping, harvesting nuts, Christmas decorations, and acquiring blisters!

Happy wooding.

4

Guests and Visitors

One of the joys of owning or caring for a woodland is welcoming visitors, showing them around, and sharing your plans and hopes for it. Many will be amused and bemused, if my experience is anything to go by! Why on earth did you buy a wood, what do you want it for, what do you do with it, how much work is it? Such questions pepper the conversation as you stroll from gate to ride to glade to the old oak to the thick stand of firs and back again. Sooner or later the question of making money arises if you have a forester with you, or its value as a pension, or hedge against inflation or inheritance tax if your friend is a city type! But all these reactions betray a curiosity, a touch of envy – at least I like to think so. Once your friends learn of your new acquisition you can expect a steady stream of visitors, always assuming you are happy to share the wood in this way.

Before looking at invited guests, we must mention the uninvited since every wood suffers trespass to a greater or lesser extent from those who have no right to be there. I will begin with a few comments about these first, so as to be rid of them before getting on to the fun bit.

Uninvited visitors

Because a wood is usually remote in the sense of not being near where you live and you can't watch over it easily, and because they are wonderful places where you aren't readily disturbed – that may be why you have bought yours – they tend to attract the occasional uninvited visitor. It's rarely a big problem, more just something to be aware of.

Casual trespass and theft
Most woodland entrances will be used by people to relieve themselves and, in a way, that is a facility the owner provides

for the desperate – we've all, surely, had to respond to an urgent call of nature! More annoying is the person who hops over the gate to help themselves to some 'free' firewood, whether from scavenging or taking from a log pile, which is why the latter is best hidden from the entrance. Some people will go into the countryside to help themselves to beanpoles and pea sticks or similar rustic items. All these trivial thefts are a nuisance though occasionally there may be a health and safety concern.

As an aside, these problems might increase. Since 'Right to Roam' can be exercised over large tracts of hillside, many people imagine that all countryside, including woodland, is now open access. Unless a wood was previously 'common land' no such right exists. So if a visitor is encountered you can politely put them right.

More significant theft may be of Christmas trees, holly, foliage, moss and other commodities used in floristry. Theft of logs when doing thinning or felling is rare, but kids climbing on stacks is not. A notice warning of the danger helps but may not stop children and may suggest to trespassers that they have a right to be there. Theft of, or disturbance to, grey squirrel hoppers, traps and other paraphernalia of pest control is a real nuisance, as are loss of tools and gear left in the wood overnight, or breaking into a shed. These problems are minimised by tidying up after work, taking home whenever possible, and disguising anything you must leave – I have an aluminium step ladder hidden behind a bush which has never been disturbed as far as I know: as I write, it was still there last week!

Keep gates secure, but remember everyone's padlock must interlink!

Other deterrents to trespass are a tidy, well-tended entrance with no litter, securely locked gates, and a freshly painted sign. Good neighbours are a blessing; so do encourage them to visit your woodland whenever they want to if they are happy to look around. My nearest neighbour walks his dog most days, or rather he used to. Mungo, his terrier, was 'stolen' we think by a passing vehicle who happened to see him and the ever-friendly dog must have happily jumped on board: he has never been found. Another neighbour who owns a small country estate a mile down the lane from my wood checks the banks and verges weekly to gather up litter – he always finds enough lager or cider cans to match the days of the working week since he last checked, but has never seen the culprit!

Poaching

One perhaps shouldn't write about poaching and wildlife crime in a book about getting started and enjoying your own wood, but it's best to be realistic. Such crime is unlikely to be met with, but still occurs. In a nutshell, poaching, whether by day or night, is unlawful. Entering land to pursue or kill game (deer, hares, several species of bird) is illegal, both in the unlawful entry and in search and pursuit with or without a gun. The offence is more serious if trespass is by several persons together, or is at night. You'll be pleased to know it is an offence to offer violence to an owner, gamekeeper or their assistants! Most police forces have a wildlife crime unit or a liaison officer responsible for dealing with such crime.

Legally taking game is restricted by close seasons that vary from species to species. And, personally, I think it is a good thing that no game may be hunted or shot on a Sunday or on Christmas day.

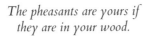
The pheasants are yours if they are in your wood.

Other wildlife crime

Special measures are in place to protect bats, badgers, certain other mammals, rare birds, plants and flowers, special habitats and so on. They are covered by the Wildlife and Countryside Act though some wildlife is the subject of a specific act. Cruelty to animals has long been an offence and even when carrying out pest control, there are restrictions on how this may be done to ensure it is as humane as possible.

Rubbish and fly-tipping

Few owners whose wood fronts a highway escape this curse. A well-tended entrance is the best means of avoidance, but it doesn't prevent it altogether, as my litany of numerous condoms, many magazines, occasional garden or builder's waste, 18 fire extinguishers (they were dumped then removed a fortnight later!), three computers, two washing machines, two burnt-out cars and one stolen gate testify! But that's over 30 years; on most visits to the wood nothing at all is encountered.

If rubbish is dumped or fly-tipping occurs, contact you local authority. The Environment Agency investigate larger scale incidents of fly tipping (generally those involving more than a lorry/tipper load of waste), incidents involving hazardous waste, and organised gangs of fly-tippers with a greater risk to human health and damage to the environment.[1] The police will probably be interested in a car that's been dumped, especially if it's burnt i.e. to destroy evidence.

Very welcome visitors

If you are as thrilled and excited by your wood as I was when I bought mine, you will be itching to show everyone your new purchase. A bit like a new house, it's a purchase you can show family and friends around. They can experience it, enjoy it, and share it with you in ways that almost nothing else equals. And that's not all. What about inviting local school children, or a natural history society, or holding a Sunday school picnic, or even an open day with a guided tour? But in all the fun and excitement you'll want to make sure that accidents are rare and that you are

---

1   If you need to report anything serious, they have a 24 hour incident hotline on 0800 807060. There are more details on their website www.environment-agency.gov.uk.

covered for unforeseen hurt in the litigious times in which we live.

Duty of care and public liability insurance

Whether access is authorised or not, a landowner or manager has a duty of care both to control it and to make risky features safe. With the earlier example of kids climbing stacks of wood, it is your responsibility to make it as safe as possible and comply with good practice. 'Managing Visitor Safety in the Countryside' provides excellent advice.[2] Also helpful is the National Tree Safety Group's '*Common sense risk management of trees*' which is available in hard copy from the Forestry Commission Publications Service or as a PDF from the NTSG website.[3] If you've invited a party to your wood it's best to carry out a formal 'risk assessment'

Do take out public liability insurance to cover claims from unwitting accidents. If you find it hard to locate an insurer, you should be able to get advice from a number of different organisations such as the Smallwoods Association, Country Land and Business Association, National Farmers Union and others. Some organisations arrange group cover for all owners with small woods. For reference my annual premium comes to about £170 on cover for claims up to £10 million. This insurance may not cover you if the access or use is charged for, and may not do so if the wood is neglected with many leaning, hung–up trees or other self-evident hazards. A bit like failing to maintain your car in a roadworthy condition, a claim could be refused if you haven't taken reasonable steps to make things safe.

Visitor access

Most visitors will arrive by car, so having somewhere to park off the public highway is a help. Inside a wood one can often park cars safely to the side of tracks, provided the ground is reasonably well-drained and it hasn't rained heavily in the last week. Of course you may need to liaise with neighbours if you have a shared access. A local farmer may be happy for you to use the corner of a meadow. In my own wood I can pack about 18 cars just inside the entrance, any more than that and my kind neighbour provides an overflow.

2 www.vscg.co.uk

3 www.ntsg.org.uk

All woods need at least one track and the secret to keeping tracks dry, in addition to being well drained, is for them to be open and not crowded in by trees. Glades beside the track, turning bays for vehicles, and an adjacent campsite or play area, all let in the sun and encourage airflow so the surface dries better: it's the same principle as hanging clothes out to dry on the line. The track surface itself can be mown grass or specially made up with chippings. If it has already been built for timber extraction it is likely to have a sound surface.

We mentioned before that the elderly need to know it's safe for them, in particular for paths to be even under foot so they won't trip or fall, and to have somewhere to sit. The elderly and the general public will also be anxious about access to a toilet: where this can't be provided simply keep the visit short. Do warn them in advance about any lack of facilities.

Most kids climb trees, make camps, gather sticks, and get stung as they muck about, so a first aid kit is essential and a mobile phone desirable assuming you get reception. The same kids, if they are local, are sure to know which mobile networks do work! Do everything to encourage kids to enjoy your wood.

Just occasionally there may be specific things to warn visitors about. In my wood in the summer there are always ticks that get on to your clothes and then creep to a soft juicy part of your body, only to reveal their presence a couple of days later. There is a slight discomfort and a 'new' red freckle with a dark centre, namely the tick burrowing into your skin! As well as fever, the nasty Lyme's disease is a risk though I've not had either. In general snakes are rarely encountered. That said, it is probably best for children to wear shoes rather than sandals and for you to know where the nearest hospital A&E department is, and not only because of the faint risk of snake bites. Some berries are poisonous like those of yew and, of course, deadly nightshade. There is also the question of mushrooms and fungi and which are or are not safe to eat.

Camping, caravans and fires
No permission is needed for occasional camping and caravanning. In the case of the latter you can keep a caravan on site for maximum of 28 days in any one year, so unfortunately your wood can't be a new home for the family trailer!

Overnighting at least once in your own wood is a must.

If cooking by open fire appeals and you want to enjoy a good old sing-song in the gloaming, do take care with siting the fire.

- It should be at least 8m from the base of any tree.
- Whenever possible try and have open sky above the fire and not branches however high above it.
- Once you have found a good fire site, try to use the same one in the future; don't move around the wood with fires here and there.
- Always make sure that a fire is well extinguished, dowsing with water if necessary, before you leave. Forest fires are rare in woodlands in Britain except in densely packed young conifer plantations or where trees are in thick grass that is dead and dry in springtime.
- If you are into Ray Mears and 'Survival' and want to start your fire with one match, remember there is always dry material inside a holly bush, whilst the outside bark of birch will peal off like tissue paper. Over this fine dry material,

create a cone-shaped pile beginning with fine twigs, then thicker ones and then small sticks. Once lit, continuously tend the fire by adding more sticks. Only once a fire is really going should you start adding split logs. With the latter use dead wood preferably not in ground contact. Even if it's raining, you will find the wood will be dry when you peel the bark off. If you have to use freshly felled wood, ash is much the best and burns quite well straight from the tree.

Invited parties

I've had many groups visit my wood. The key to a successful visit is to plan where you will take the party and what you will show them that is of interest – what interests you will interest others. I plan for about 8-10 short stops which might take about an hour or hour-and-a-half to go round. Groups up to about 20 are a good size; more than this and the meandering crocodile takes a long time to assemble at each point and, of course, you have to raise your voice, especially on a windy day. Groups of more than 30 are best shown around by laying out a self-guided route, but this takes time and effort.

A visit to a woodland can bring alive any nature class for youngsters or interested teenagers doing biology. If possible show the teacher around first to assess the hazards – and so help with the school's risk assessment, and also to see how the visit will fit with the day's lesson plan. As Alex Argyropulo, one of the original book's reviewers, said: "It is sheer delight [to run a forest school] but one really has to return the children to parents in one piece!"

Open Days

I mention this topic because I have now run several and they are greatly enjoyed. The idea began following publication of *A Wood of Our Own* when people began to ask if they could visit the wood in the story. So rather than ones or twos being shown around we decided to hold an Open Day with a laid out route of 12 stops and invited about 150 people – family, friends, folk from church and, of course, those who had specifically got in touch. It worked well and so far we have run seven such events. May Day or Spring Bank holiday weekends are good times with spring flowers at their best.

Working 'bees'

I'm always surprised how keen people are to come and do woodland work for weekend relaxation. Groups like British Trust for Conservation Volunteers (BTCV) and local natural history societies have long known this. So if you have jobs needing doing – cutting a hedge, planting trees, coppicing, tidying up a track, gathering firewood – you will probably be swamped with offers once the word gets out! Do make sure that everyone comes well equipped for the conditions, that hazards are pointed out, tools are sharp, and any chainsaw is only used by a fully qualified operator. Indeed, it's probably best to stick with bow saws and other hand tools.

For us, amongst many sorts of assistance, our church youth group have helped excavate a pond, our pastor has high pruned several trees and helped cut back hazel, the local doctor's family have spent several days coppicing. My Imperial College students would try their hand at thinning, pruning, stacking cords of firewood, clearing scrub, coppicing, burning lop and top – in fact having a whale of a time – but under strict supervision.

No payment should be made or else you get into the realms of entering into a contract with all the health and safety, insurance and employer liability implications this brings. Of course, a lovely meal at the end of the day, or toasted marshmallows over an open fire – the choice of our church youth group(!) – will be appreciated, but even this should not be presented as payment in kind, otherwise you could be deemed to be their employer.

Pest control, shooting, hunting

Sometimes a small wood may be part of a larger area where neighbouring landowners are keen to control pests, cull deer or shoot game. There is no requirement for you to give permission to enter your land. However rough shooting – rabbits, grey squirrels, wood pigeons, etc. – can be offered to a local group who may visit a couple of times a year to keep these pests under control. As we remarked in an earlier chapter, you could well end up with a haunch of venison in exchange for allowing a couple of hours of hunting on your land to help with the enormous and much needed task of deer control!

When you invite or allow such people into your wood, do check that they are properly qualified, have suitable insurance, a gun licence, and comply with the law concerning close seasons. For further information about sporting use, contact the British Association for Shooting and Conservation.[4]

As William Cobbett said
in Rural Rides*:*
'What in vegetable creation
is so delightful as the bed
of coppice bespangled with
primroses and bluebells?'

---

4   www.basc.org.uk

5

When You May Need Permission

This threatening title is meant to be nothing of the sort; it's a reminder that there are a few things you can't do without first obtaining permission. It's not a chore because most of the time officialdom and bureaucracy simply don't figure in how you run your wood, but like planning permissions and building regulations for your house, there are some things about which the state or local authority want to have a say.

Planning permission

In general most woodland work and operations do not come under the normal planning regulations: the major exceptions are if you want to build a house or create a new access on to a public highway.

Many of us like the idea of living in the country and what better place than to live in your own wood! This prospect, unless there is already a dwelling, is well-nigh impossible to achieve. Buying a wood and then seeking planning permission to build a house will fall foul of virtually every planning authority's guidelines, but see footnotes on pages 16 and 22.

One possible exception is where your woodland is large and you can make a credible case that the wood is your main source of income or livelihood. You can then apply to build a forest worker's dwelling, which must be appropriate to the job – not a six bedroom four bathroom country home. Moreover such a cottage or small house must continue to be used for this purpose: it cannot be sold a few years later say for ordinary residential use without expressly applying to the planning authorities for a change of use. Such a change is unlikely to be granted.

What you can do in your wood by way of accommodation is camp there or site a caravan, as we mentioned in the last chapter, for up to 28 days. Permitted development also includes a secure tool shed, provided it is sited well away from a highway, but even here one needs to be a little careful. A shed must be a shed – a small hut with no windows and a lot of cobwebs – not something that can double up as a summerhouse or a modest chalet!

Felling trees

In general you can cut and fell small trees, but when it comes to several or more big ones you will almost certainly need to obtain a felling licence. I shall try to explain the basics in simple terms. I am assuming there is no tree preservation order (which is very unlikely for a wood unless you are adjacent to a built up area) and that your wood doesn't have special status such as a National (or local) Nature Reserve, a Site of Special Scientific Interest (or other official designation), and is not in a conservation area.

The felling of trees in Great Britain is controlled by the Forestry Commission. Now here is the official bit: A *licence is normally required if you want to fell more than 5 cubic metres of timber for your own use in any 3 month period or just 2 cubic metres if it is to be sold.* What does this mean?

A tree with a trunk containing one cubic metre of wood is quite a big one. For a conifer, such as pine or spruce, it will be a tree you can just about hug and get your arms around. Its diameter at chest height – strictly when measured at a point 1.3 metres from the ground – is around 40 centimetres. For broadleaves, such as ash, oak and beech, the same applies but their trunks and crowns tend to be less uniform so this guide is a bit more rough and ready. That said, you can see that if you only want to fell a few trees over a period of time, say to develop a picnic glade, a licence will not be needed. More than this, and it will be. Felling licences are not difficult to obtain and are not withheld without a really good reason. Contact your local Forestry Commission (or Natural Resources Wales) and ask to speak to your local 'Woodland Officer'. The woodland officer is responsible for all of the grants and licenses issued by the Forestry Commission in your area and will be able to advise you whether you need a licence and how to get one – use the internet or via Yellow Pages for the phone number.

There are some exceptions when you do not need to apply for a licence:

- If the felling is part of an already approved plan of operations by the Forestry Commission
- The trees are in a garden, orchard, churchyard or public place
- The trees are small less than 8cm in diameter at breast height, less than 10cm if part of a thinning, or less than 15cm if the material you are cutting is coppice or underwood e.g. hazel or mixed scrub
- Trees that are dead, obviously dangerous or are nuisance (note that in a conservation area you cannot fell dead trees without permission).

Where clearfelling is intended, that is an area of your woodland is all cut at one time, the granting of a felling licence invariably includes the condition to restock the land with trees, by natural regeneration, coppicing or by replanting, as appropriate. Clearfelling woodland and turning land over to another use is only allowed in exceptional circumstances. Grant aid may be available to help with regeneration and woodland improvement (page 135).

European protected species

European protected species are animals and plants that receive protection under the Conservation of Habitats and Species Regulations (2010) in addition to the Wildlife and Countryside Act (1981). The species itself and both where it breeds and its resting places are protected.

As well as being an offence to deliberately capture, injure or kill any such animal or their eggs, it is an offence to damage or destroy a breeding or resting place. Tellingly, 'deliberate' is taken to mean both 'intentionally' and 'recklessly'. This definition covers all forestry operations, whether you knew the species was present or not.

When working in a wood, follow best practice as far as you can. This may mean doing a job at a less sensitive time year to avoid nesting or hibernating, modifying the scale of an operation, or deliberately setting aside a part for conservation. There is a huge amount of advice some of which we refer to later in Chapter 9. The main protected species likely to occur in woodlands are:

- All 17 species of bats
- Dormouse
- Great crested newt
- Otter
- Sand lizard
- Smooth snake
- The natterjack toad and some plants such as marsh saxifrage may rarely occur in woodlands or be affected by forest operations.

The Forestry Commission has extensive guidelines for dealing with all these species that can be accessed on line or from you local forest office. When in doubt, don't panic, but do get in touch with your local woodland officer who should be able to advise you.

Planting trees

You do not need permission to plant trees. If you are seeking grant aid to plant, conditions may apply such as providing public access, planting only native species on ancient woodland sites, and so on. Woodland grant schemes differ in detail between England, Scotland and Wales, but all seek to support the owner to provide public benefits in exchange for receipt of public funds.

Creating ponds

While building dams, reservoirs or ponds may not be uppermost in your first thoughts about looking after a wood, it is worth noting that diverting a watercourse to make a pond or abstracting water from a borehole requires permission under the Water Resources Act. Small rain-fed or groundwater ponds are unlikely to be affected. We look at creating a wildlife pond later in Chapter 9, pages 111-112.

Easements

This is a term in property law where rights exist over another's land. For example you may have the right to use a track across a neighbour's land, the electricity board may have purchased the right to take electricity by underground cable or with poles and wire across your land, or a third party may have right of access along your track to gain access to some facility like a mobile phone mast, etc. These rights rarely intrude in your enjoyment of the wood, but if what you plan to do interferes with their free exercise, obtaining agreement for the temporary restriction is advisable. For example, logs cut from thinning your trees need to be stacked and this could lead to blocking the track or making it impassable for a few weeks or even months.

Public rights-of-way

When you purchase a woodland, as mentioned in Chapter 2, the existence of rights-of-way should be evident. Obviously they must be kept open. Indeed, taking care to keep them in good shape greatly reduces the chance of members of public wandering off and going where they are less welcome. All of us prefer to stick to a good clear path, so maintenance benefits the public and owner alike. Varying the route is a long, but not impossible process, provided a good case can be made. Your first point of call will need to be with the County Planning Department who hold definitive maps numbering and classifying every public right-of-way. The planning people at your local district council office will also be worth contacting.

You should also bear in mind that if paths on site have been subject to unchallenged use by the public for at least 20 years there may be a 'presumption of dedication' under Section 31 of the Highways Act 1980. This means they have become rights of way be default.

Wildlife management

We touched on this briefly in Chapter 3, but if you intend to shoot you will need a gun licence and, as mentioned, you must comply with the law concerning close seasons. In brief rabbits can be controlled at any time, grey squirrels at any time unless poisoning with warfarin (which may be withdrawn) that is restricted to 15th March to 15th August, and deer usually in the winter, except for roe males, but specific dates apply to each species and each sex, so do check.

Poisons, such as phostoxin to control rabbits, will require your signing the Poisons Register at the time of purchase and keeping them in a secure place.

Interestingly, it is the landowner's responsibility to keep down vermin so as not to be a nuisance which, from a forestry point of view, includes rabbits and grey squirrels.

A final word about permissions

We have covered the main situations where permission may be needed, but other activities where officialdom will be interested would include: holding car and motorcycle rallies; paintballing and other pursuits where charges are made; erecting advertising hoardings beside a highway; and any work on or near a site of archaeological importance.

This chapter hasn't been so bad! You are remarkably free to look after your wood how you want to. In my 30 years of ownership, I've not felt hemmed in by bureaucracy or frustrated by red tape; I hope you won't either. Let's hurry on to the action, I haven't even found a suitable illustration that might delay us further, let alone amuse or inform.

6

Growing and Caring for Trees

We British are great tree planters. Our woefully denuded countryside was down to 5% tree cover 100 years ago, now it is more than two and half times that thanks to planting. Our success and experience makes us a world leader in sympathetic planting – virtually all our state forests are certified[1] – though you wouldn't have said that 50 years ago when we were coniferising ancient woodlands and carpeting hills with square blocks of dark green spruce! Our history of planting goes back many centuries: we have all surely heard how in the 17th and 18th centuries the cry was to plant oaks because the navy was running short of shipbuilding timber.

This is a long chapter and I've divided it into four sections to look at:

- How to create a plantation;
- How to encourage trees to develop from natural seeding – what foresters call natural regeneration;
- What may need doing as a stand of trees develops; and
- Coppicing and pollarding.

PLANTING A WOOD

Planting trees - the basics

Where to plant trees
An odd question to begin with, but are there spots or places to avoid? I ask this first because visitors to my wood who see gaps, from a recent thinning or creation of a glade, often assume that tree planting is now needed: it isn't. Generally we are overly

---

1 Certifying of forests first began about 20 years ago and is an independent vetting that management meets agreed standards of good practice to ensure sustainability.

keen to fill up ground. Remember that existing trees need space to grow and wildlife thrives in gaps and glades. More specifically do not plant trees that might damage valuable habitat such as wetland or right next to streams and, similarly, not right beside rides and roads. If you do plant up a large opening, only plant trees in the middle where there is clear sky overhead; young trees don't grow well in the shade of old ones.

So, plant trees to turn a field into a wood or as a way of regenerating a large patch that has been clear felled, otherwise think hard about whether you really need to.

As an aside, if you need to add trees to a hedgerow, try recruiting a sapling by not cutting it when hedge-trimming. Make sure the sapling is growing directly above where it is rooting to avoid developing an 'S' bend and remember to mark the sapling with a stick or ribbon so whoever is trimming the hedge can see it easily. Will recommends a scaffolding pole, but do make sure it's clearly visible!

What species to plant

In Britain we have three native conifers and about 30 native broadleaves. Introductions from across the world in the last 300 years have immeasurably added to the possibilities. In the appendix there is a list of the commoner species, what they are good for and the conditions that best suit them. If in doubt consult a local forester, usually Forestry Commission or Natural Resource Wales staff will be happy to make suggestions, and do note what is growing well already in or near your wood. You will rarely go badly wrong by continuing to grow what is doing well in the vicinity.

A question of provenance

Provenance is a special forestry term that tells you something about where the seed for your plants (young trees) came from. It is important because if you are planting oak seedlings grown from acorns collected in Scotland you would expect them to grow a little differently compared with ones imported, say, from France. Each area has developed some genetic traits: ones from Scotland may grow more slowly but be hardier to late frost because they flush later in May than ones from France. Indeed, in general in Britain, we find that trees of more southerly provenances grow quicker but are more tender than northerly ones. This is just one example, but may be an important one as climate change increasingly takes hold.

The topic is complex and is the subject of much scientific research. Again, your local woodland officer or any qualified forester will be able to give good advice. You are unlikely to go far wrong by planting trees raised from a local seed source. Sadly over the years Britain has imported much seed from unsuitable places e.g. oak and beech from southeast Europe simply because, in their case, home supplies were unavailable. Both species are erratic in seed production, only having good years, 'mast', at long intervals.

Where can small trees be bought?

Contact a tree or forest nursery, not a garden centre. There are many nurseries and your nearest will not be too far away. In the Yellow Pages they are entered under 'Nursery Horticultural – wholesale' and/or 'retail'. When choosing a nursery make sure the advertisement mentions trees, then find out if they have what you want and for how much. Most nurseries like orders well in advance and then they lift the plants you want – 'plants' are what the trade calls small trees suitable for planting – and dispatch them to you just before the date they are to be planted. Sometimes plants will be lifted in November or December, carefully packed in bundles in purpose-made double skinned polythene bags, and then kept in a cold store for dispatch in late winter or early spring. The best thing is to contact your local nursery as early as possible in the autumn to discuss your requirements.

If you like the idea of developing your own small tree nursery there is plenty of advice in books. We will look at the subject briefly a little later.

What size of plants are best?

Usually it is the smaller the better. A plant 20-30cm tall and 5-7mm thick at the root collar i.e. at the point where stem and roots meet, will readily become established and start growing well once the weather warms up, provided of course it is protected from browsing damage and vigorous weeds.

Bigger trees, such as whips, saplings, half-standards (3m tall) and standards (more than 3m tall), all impose increasing strain on the newly inserted root system. Their crowns of leaves demand many litres of water each day which the roots cannot easily deliver because many were broken off, especially the fine roots, when the trees were lifted from the nursery and because they haven't had much chance to re-grow in the new soil. Not surprisingly, large trees planted for effect on housing estates can languish for years with thin crowns and poor leaves. Worse, large trees are 100 times more expensive, and I'm not exaggerating. It may seem counterintuitive, but the shock to larger trees of being moved is so great that they often reach a mature size well <u>after</u> trees that have been planted as small transplants or bare-rooted whips.

What kind of tree to plant?

Seedlings and transplants are the two commonest names you will come across in a nursery catalogue. When they are sold with no soil attached to the roots, they are called 'bare-rooted' plants. We will define them along with some other types of plants.

Seedling – This is a plant that has grown up from seed, but has never been moved in the nursery until it is lifted for dispatch after it has been sold. Seedlings are usually one or two years old and 15-50cm tall.

Transplant – This plant began life as a seedling but was moved, when dormant, from one nursery bed to another, hence the name 'transplant'. The moving between beds severs taproots and helps ensure a stocky, more robust, plant with a compact root ball. They are the commonest type of plant used and are mostly 2-4 years old and 15-75cm tall.

Undercut – This plant is a seedling that has had its roots undercut in situ at a depth of 10 or 15cm to achieve the same effect as transplanting but without moving it from one bed to another. They are usually 1-3 years old and 15-75cm tall.

Whips – Plants that are allowed to grow beyond the seedling/transplant stage for another one or two years in the nursery to reach 80-150cm tall. If well furnished with branches, they are known as 'feathered whips'.

Some of the types of plants/trees for sale at a forest nursery (<u>not to scale</u>):

(i) transplants *(ii) container grown tree*

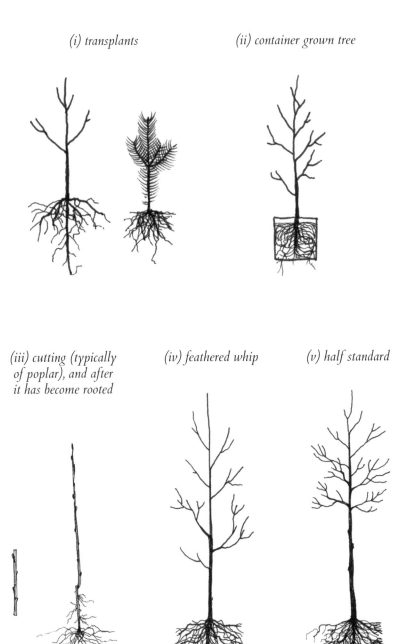

(iii) cutting (typically of poplar), and after it has become rooted *(iv) feathered whip* *(v) half standard*

In a nursery catalogue the type of plant and its age is indicated by a simple code. A seedling is just given an age. A transplant may be shown as 1+1 meaning it is two years old, having grown one year as a seedling and then been transplanted to another bed for a further year. If it grew for two years in a transplant bed the catalogue shows this 3-year-old plant as 1+2. An undercut is a 1u1 if it grows as a seedling for one year, was then undercut, usually in June of the second year, before being sold. Sometimes undercutting is done during the first season in July and one year old plants and are shown as ½u½, but are seldom sold.

Container-grown stock
Growing small trees in containers is quite common. It gives each plant a root ball and is used for trees, such as birch and Corsican pine, that lack fibrous roots to improve establishment. In general, forest trees are bought as bare-rooted stock. There are many kinds of containers and many names for them: containerised stock, Japanese paper pots, cell-grown stock, plant plugs raised in pre-formed plastic trays, pot grown stock, and ones raised in a hinged 'book' such as 'rootrainers'. Container grown trees are more expensive, typically twice the cost per plant of bare-rooted ones. Many nurseries do not stock all tree species in this form.

Plant quality
So far we have considered what type, what size and from where plants can be obtained, but what about quality? It is important for plants to be sturdy to survive the rigours of planting whatever the site. Although height of plants is commonly given in catalogues, root collar diameter i.e. how thick it is at the point where stem and roots meet, is more important. Research shows unequivocally that thickness of stem, not how tall a plant is, is the most reliable indicator for subsequent survival after planting. Basically one wants healthy plants that have thick root collars, not weak flimsy ones. Equally important is a vigorous, fibrous root system, not a single long taproot.

The plants themselves must be free of damage, disease and other defects.

When plants arrive

Forest plants (young trees) are dispatched in the special double skinned polythene bags that are black on the inside, sealed and designed to minimise heating if ever exposed to the sun, even if briefly. On receipt of your plants, open the bag and check that the right species have been sent and that the specifications are as ordered. Then take one or two plants from a few of the bundles – transplants are typically in bundles of 25 or 50 – and check their health by nicking the bark on the stem with your thumbnail. The tissue beneath should be green or greenish-white. If it is at all brownish then the plant is likely to be dead even if the buds and other parts look OK. If more than one or two are brownish, reject the entire consignment and complain to the nurseryman.

Once you are happy with the plants – they are what you ordered and appear healthy – gently retie the bundle and close the bag. Store the bag(s) in a cool, dry place, preferably in the dark or in heavy shade, and make sure it is frost-free. An unheated garage is ideal provided it never cools below freezing. While dormant in winter and early spring, such plants will remain alive for many weeks. Normally plants will stay damp enough in the bag they came in if it is kept sealed until time for planting. Don't let the tree's roots dry out for any reason, as this is likely to kill it before you've even planted it in the ground.

Try not to move the plants until the day of planting. In particular DO NOT manhandle the bags, NEVER throw them around or drop them: treat them like your finest dinner service! The fine roots of young trees are easily broken and damaged, but this is rarely obvious to see. Experiments by Forest Research (the research arm of the Forestry Commission) show that bags with bundles of plants that are dropped, even from a height of only one metre, reduced the survival of the young plants when planted out. The more frequent the dropping and mishandling the worse survival and growth became. The take home message is: handle with care.

Obtaining trees for planting need not be an expensive business. A little forethought and care of the plants received is all that is required.

When is the best time to plant?

Plant trees anytime from mid-autumn to mid-spring, but avoid very cold and frosty weather. In the south of Britain all planting should be done by the end of March, in the north by mid-April. Trees must be planted while they are dormant and <u>before</u> they start coming into leaf (flushing).

How far apart should trees be planted?

Usually trees are planted 2–3m apart and equidistant along and between rows i.e. square spacing. You mark the position with a stick, or if planting only a few trees judge the distance approximately by pacing. There's no need to lay out a new wood or a belt of trees with the 'theodolite' precision seen in many an orchard.

If you are planting a thick hedge, three rows are a good idea and use triangular spacing. Set the trees about 1m apart.

How is planting actually done?

While a nicely dug pit is best, a simple slit in the ground with a spade or mattock is often sufficient. Gently place the roots in the slit or pit and tease them out. Then, while holding the tiny tree vertical, replace soil and firm it home with the ball of the foot. Make sure that the tree is planted only to the depth of the root collar, neither deeper nor shallower.

John White's idea of my brother-in-law and me planting trees in my wood – plus fours are not de rigueur.

It is sometimes recommended that broken or damaged roots are trimmed with secateurs before planting. This takes time, is probably of little benefit and the further handling of the tender root system will break off more fine roots. The only time trimming will help is where roots are long and straggly since it's no good simply stuffing everything into the ground.

Do I need to stake my trees?

No is the short answer, unless the trees are over 2m tall – which is not recommended. If staking has to be done, use a <u>short</u> one and secure the tree at about 30cm from the ground. Do not fasten a tree to a stake at 1.2-1.5m; the tree needs to flex from the base to build its stem properly. All a stake does is to help anchor the root system. Do remember to loosen the tie each year. By the third year a stake should no longer be needed.

Do trees need fertilising and irrigating?

In general, 'no' to both questions. If tree planting is done properly neither fertiliser nor irrigation will be needed. Phosphate (P) may help on some poor upland sites or sandy soils, but elsewhere the soil is already fertile enough for the undemanding appetites of trees. If you've planted the trees properly in the autumn or early spring and provided competing weeds are under control, you shouldn't need to water them in the summer however hot and dry it gets. Only large trees such as half-standards may need irrigating just to keep them alive, but you know what I think about such planting stock already!

A summary

The key points are:

- Generally use small, thick-stemmed plants bought from a reputable forest nursery;
- Make sure plants are alive and healthy – a thumbnail nick of the bark should reveal greenish tissue;
- Always keep plants cool, shaded and prevent roots from drying out – treat them gently;
- Plant trees while still dormant and before starting to flush i.e. by the end of March;
- Plant during cool, damp weather and not when it is frosty

or very dry;
- Plant the tree by inserting roots carefully into a small pit or slit in the ground and gently firm the soil with the ball of the foot.

What happens next?

So, after the planting, what then happens? Put another way, how are newly planted trees looked after during their first few months and years? There are really only two main concerns: protect them from damage, and don't let weeds get the upper hand.

What threatens young trees?
Young trees are vulnerable to several dangers. Particularly damaging are browsing animals, such as rabbits, hares, deer or indeed livestock. If they are present, newly planted trees must be protected. Some of these animals will also gnaw or strip bark.

Protection from late spring frost, drought or strong winds are largely outside one's control. Overcoming frost and drought problems is mainly resolved by choosing hardy species to plant and, for example, not planting in a frost pocket or on thin soils that readily dry out. Protection from insect pests and diseases is usually not needed. There is a nasty weevil that costs owners of commercial forests a lot of money to control, called *Hylobius abietis*. It has a predilection for the bark of very young conifers (and other species), but it usually isn't worth doing anything about it when planting only a few trees and, anyway, on many sites it doesn't cause any damage at all.

How do I stop animals causing damage?
The main choice is between fencing or individual tree protection. For individual protection there are many proprietary products to choose from. All are variations on a theme of plastic tubes or plastic mesh. The main consideration is the browsing animal itself and the height it can reach. For rabbits and hares a 60cm high guard is sufficient, for roe deer 1.2m, and for red or fallow deer 1.8m. Very roughly, use individual guards if you need to protect fewer than 1,000 trees, but consider fencing if you're planting more than this in one block. The basic specifications

for a rabbit and deer proof fence are sketched here. Typically a contractor will charge around £10 per running metre to erect.

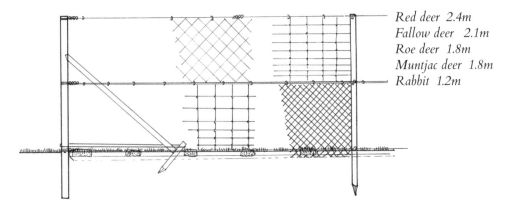

Red deer 2.4m
Fallow deer 2.1m
Roe deer 1.8m
Muntjac deer 1.8m
Rabbit 1.2m

Rabbit and deer proof fence showing typical specifications.
You will only need this if you are into serious planting.
For small numbers use individual tree guards.

Smearing repellents on the growing shoot of a tree is a nice idea though the only useful product is called *Aaprotect*. It lasts for a few months but only affords protection to the parts of the tree actually smeared. Synthetic lion dung and other seemingly fearsome smells have been much hyped and on the whole, found much wanting!

I am going to insert here an aside about a topic that doesn't fit happily elsewhere in this small book: the control of an animal that is the scourge of broadleaved woodlands – the grey squirrel. This menace from North America strips bark from oak, beech and sycamore trees (and many other species) and will attack trees as small as the thickness of one's arm to nearly mature ones. Damage occurs mainly in May, June and July. If uncontrolled they can ruin a promising stand of trees disfiguring and sometimes killing them.

Grey squirrels have been designated a pest since the 1930s. They are controlled by poking their nests (dreys) and shooting animals that emerge, by trapping them, or by poisoning with warfarin dispensed in specially designed hoppers (though at the time of writing there is uncertainty whether continued use of warfarin will be permitted). The law is strict about what can and cannot be done. Even if you like grey squirrels in your garden, they are no friend in your wood. They damage your trees, probably raid birds nests for eggs, and may harm other wildlife.

Now back to your newly planted trees.

When should such protection start?
On the day you plant your trees! All too often planting has been done one day with the intention of installing protection the next only to find devastation occurs that very night. Animals familiar with a territory soon know that new and succulent plants have arrived – there is some evidence that freshly planted material actually emits chemicals that can be sensed. Some years ago I planted six fine young cherry trees in my wood and I thought I could get away without protecting them. Within 24 hours each was sought out and nibbled or bark stripped. Don't wait for damage before giving protection.

Do I need to worry about competing weeds?
Yes, this is important since weeds compete with young trees for moisture and nutrients and, if they grow tall, for light as well. Some weeds such as wild clematis and wild honeysuckle will also entwine and even overwhelm saplings.

How are weeds controlled?
Weeds are best controlled by killing them either with herbicide, by applying a mulch (a material placed on top of weeds to suppress them), or by hoeing around the tree. Just cutting weeds is ineffectual. Many owners of small woods will not want to or really need to use herbicide, particularly if you have willing labour to hand! If you do go down this route, use a proprietary herbicide, but take great care to avoid any of the herbicide getting on to the tree. If weeds were removed before planting, one application of herbicide in late June or early July will usually be sufficient to kill weed regrowth that has developed. Do read the label for instructions and follow them to the letter. Herbicide can be applied either with a directed spray or wiped on with a device that has a wick moist with the herbicide.

A good alternative to herbicide, particularly if you only have a few trees is to use mulch. Many materials are suitable, such as bark chips, black plastic, and even old carpet placed around the tree. Make sure that inert mulches like black plastic are thick enough e.g. farm silage wrap, and well pegged down. Organic mulches should be 6-10cm deep; a light mulching is useless as weeds simply grow through. One attractive combination is to use plastic sheeting as the main mulch and then sprinkle bark chippings on the surface both to keep it in place and as a disguise. Mulching is best done at the time of planting.

Cutting or mowing weeds is not very effective in relieving stress to young trees. Although cutting reduces competition for light, it usually worsens it for moisture and nutrients as new weed growth with new demands is stimulated. Just think of a lawn: for as long as it is kept green by mowing it is necessarily drawing moisture!

How far from a tree should weeds be controlled?
Control weeds to a distance of about 60-70cm from the young tree or a diameter of 1-1.5m. There is no need to kill all weeds between rows of trees, just concentrate on controlling those immediately around the tree itself.

How long will weed control be necessary?
Continue weeding until the tree is about 2m high. This will usually be for about three years, and one weeding per year, provided weeds are killed, will usually be sufficient.

Do I need to replace trees that die?
Not necessarily. Since over the life of a stand of trees one will usually thin them out at intervals, it is not necessary to replace every tree that fails in the first year or two after planting. A rule of thumb is to replace failures if more than one in five dies. Less than this, and the odd death doesn't matter.

Starting trees from scratch - Seeds, Seedlings and Tree nurseries

As we've noted, most trees start life as seeds sown in a special forest nursery where they are cared for until robust enough to withstand planting out on the site where they will spend their life. This is very different from growing corn or cultivating vegetables where a seed is usually sown in the place where the plant will grow. However, sowing seeds directly on a site to be restocked can sometimes work.

Sowing seeds directly

One can be tempted to find one's own seeds – acorns, beech nuts, seeds that fall out of a dry pine cone etc. – to sow directly into the soil; after all even 30-50p per plant from a nursery sounds expensive when wanting to plant hundreds of trees. But sowing directly is more difficult than it might appear and I list the reasons why, as a reminder of what we need to think about when establishing trees.

- You can't be sure seeds are viable and often only a few will germinate
- Some seeds such as those of ash need cold, moist stratification[2] to germinate well
- Sowing depth can be critical, neither too deep nor too shallow, but this varies between tree species
- Seeds are eaten by mice, voles, birds and squirrels and you need to sow many times more than the number of trees you hope to establish
- Tender seedlings that do emerge are easily swamped by other vegetation or may be browsed
- Seedlings from seed sown in a wood generally grow much more slowly in the early years than those from a nursery.

For all these reasons it is far, far safer to buy and plant healthy young trees, already 1–3 years old, that have been raised in a special nursery with the care this affords. They will continue to grow faster for several years than seedlings in a wood.

If you are anxious to save money at this stage, then dig up wild seedlings, called 'wildings', from elsewhere in your wood and use these. Do it properly by excavating a decent root ball and planting the tree in the new location as soon as possible. It is time consuming and, of course, one is restricted to what wildings

---

2 Stratification is the storage of seeds at just above freezing temperature in damp sand for several weeks or months to mimic winter soil conditions. This 'pre-chilling' is required for species with deeply dormant seed to improve germination when sown.

are to hand. You often find that such self-sown seedlings have astonishingly long, straggly roots that are difficult to dig up. Their lack of fibrous roots slows establishment and early growth.

Your own nursery

It can be enormous fun growing your own trees from seed. Your friends can help too, sowing acorns in flowerpots and bringing to plant in your wood as trophies and trysts of friendship! But if you want to grow your own seedlings for a significant planting you will require a site that is quite demanding in its requirements, but will often be met by the corner of the vegetable patch in your garden! The key requirements are:

- Full protection from browsing animals – voles, rabbits, hares and deer
- Full protection from seed predators – mice, birds and squirrels
- Access to irrigation
- Soil that is friable and easily worked and kept weed free
- A location that is not a frost hollow or excessively exposed
- Adequate access.

If you have such a site and want to try your hand, I attempt to do here for tree seeds the sort of instructions one finds on the side of every flower or vegetable seed packet.

The Tree Council issue an excellent booklet entitled *The Good Seed Guide – all you need to know about growing trees from seed*.[3] It is most helpful and well worth getting if you want to grow your own. It is illustrated in colour and silhouette images, many by John White who has also illustrated this book and the two telling the story of my own wood! In the booklet's information sources at the back, one source cited is John Evelyn's famous *Silva,* published over 350 years ago, and that rather took my breath away! The address of the Tree Council is in the appendix.

I will cover here the main topics in brief and content myself with strongly recommending that you obtain the Tree Council's excellent guide.

---

3 *The Good Seed Guide* published by the Tree Council in 2001. ISBN 0-904853-01-2. The Tree Council's address is 71 Newcomen Street, London SW1E 1YT Tel: 0207 407 9992 Fax: 0207 407 9908 Website: www.treecouncil.org.uk

Where to get seeds from

If you want to collect your own seeds, obtain them from several different trees of the desired species not just from one, and from trees that are healthy and of good form. You can also buy tree seeds from certain nurseries. Whatever you decide, supplies of seed from year to year are not consistent since most trees only fruit heavily at intervals.

Sowing seeds

When – Sow seeds in mid to late spring once risk of frost is past.

Condition of bed – Prepare soil into a tilth, clean away all weeds, incorporate a slow-release compound fertiliser high in phosphate (P), but don't over-fertilise.

Quantity – Aim for 100 seedlings per square metre of bed or sow two or three per container.

Depth – Sow small seeds, such as birch, on the surface and very lightly cover: larger seeds, such as acorns, should be sown to a depth of 1-1.5 times their longest dimension. They needn't be placed point upwards as they naturally germinate on their side.

Seed covering – Cover seed to prevent birds eating them and help keep them moist. Use sand or grit that is on the acid side of neutral, not chalky. Sandy loam will do.

Watering – Irrigation is critical, but mustn't be overdone. How much depends on weather conditions: the bed or container soil should neither be soggy nor dry and dusty.

Protection from mice and birds – Use netting to keep birds away in the precious days and weeks between sowing and germination: a cat might deal with both threats!

Germination

Germination of seeds never occurs uniformly owing to the seeds' condition, the nature of the seedbed, and other factors not under your control. Germination time varies between species. Expect some seeds to germinate a few days early, most to come up at about the average time and a few to be slow and only emerge later. There will be a definite peak when most come up. In the days after germination provide shade if it is at all warm and sunny since young seedlings are very tender.

Weed control
Seeds of weeds often germinate quickly and the weeds themselves often grow rapidly and soon overtop your tiny trees that must be helped at this critical phase. Pull up all weeds. Do this once a week for the first couple of months and then as conditions dictate.

Thinning and pricking out
When sowing flower or vegetable seeds we are used to thinning out seedlings to give the remainder room to grow. This is less common in tree nurseries except when using containers (you only want one good plant in each). If you are thinning down to one plant per container, plant the surplus seedlings to make good any containers that are empty. This operation, a form of pricking out, is best done when seedlings are small, 'matchstick' size. Ease the seedling from the container soil and straightaway re-pot it in the new container using a miniature dibble, such as a small stick about the thickness of your little finger with a pointed tip. No time should elapse between lifting and re-potting the seedling apart, possibly, from a brief dunking of the roots in clean water. Do the work in the shade. These precautions are essential: it doesn't matter how careful you are as lifting and transplanting will break off many fine roots – the tiny tender roots and root tips you can only see with a hand lens. Without these the seedling will struggle until they becomes re-established again in two or three days time.

Freshly pricked out and re-potted seedlings should be kept shaded, for a week but not in the dark. The amount of shade can be reduced after about three days.

Transplanting and undercutting
In a small nursery simply use an ordinary flat spade for transplanting. Do it when the plants are dormant, between late

autumn and mid-spring, but not when ground is frozen. Sever the roots at a depth of about 15cm and lever up the plant. Gently shake the soil off, place the seedling in a polythene bag in the shade, and replant it as soon a possible in its new bed for another year's growth. It is now a 'transplant'.

To undercut, use a spade sharpened with a file. Angle the spade down to cut the roots at 15cm depth, and then withdraw. Undercut in the growing season, around July.

Lifting and despatch

Your baby trees are ready for planting, usually after one or two years in the nursery when they are 15-50cm tall and 5-8mm thick at the root collar. The final job is to lift them. With a small number of plants, simply repeat the transplanting operation to extract them from the bed, gently shake off soil and place them carefully in polythene bags kept in the shade. Provided plants are dormant when lifted and are kept in a cool dry place – and not at risk from frost – they can be stored for up to two months before use.

Now follow the operations for normal planting we outlined earlier (page 54).

If all the above operations proceed as planned, you will have the immense satisfaction of growing trees from the very seed you collected. There is no surer way of being certain of their origin and perpetuating the trees and shrubs you want.

NATURAL REGENERATION

Most kinds of trees in Britain readily set fertile seed that falls or is carried by wind or birds to become the next generation of young trees. We all know someone whose lawn is seemingly invaded year after year by sycamore seedlings or baby oaks germinating from acorns, or how a patch of bare soil or a neglected corner can soon sport some healthy birch or ash. Such 'volunteer' trees are called 'natural regeneration'. In our wood this free yield is mostly the occasional birch and ash except on the south side where carpets of sycamores have arisen thanks to seeds from a couple of large, old specimens nearby.

A jay, the master of oak natural regeneration.

But are free seedlings like this always welcome and when should you utilise them? Firstly, the trees may not be growing where they are wanted and there's no point in encouraging trees to grow simply to fill up some land. You can dig up such self-sown trees and transplant them but, as mentioned earlier, you will find that even small ones have roots that will have taken a strong hold and often penetrate to a surprising depth. Experience shows however, that even with great care such 'wildings' often don't grow very well.

Secondly, not only may self-sown seedlings not be of the kind tree you want but also they are not always found growing on a site suited to them. Common examples are ash and birch which spring up in dense thickets on wasteland and many freshly exposed road and railway embankments. Despite their profusion they will not develop well on such sites and neither highway authorities nor, certainly Network Rail, will welcome them either!

Where natural regeneration comes into its own is when you want to perpetuate the local tree stock, perhaps for reasons of conservation. Using self-sown trees continues the line of the parents and this goes back generations – centuries in the case of some trees – and so is a genetic link with the woodland heritage of an area. This is mainly a concern with native species and provided the stock is adequate for your entire tree growing objectives.

Deliberately encouraging natural regeneration is tricky. As a way of regenerating woodland it is attractive, but seeds may not be available in the year you want to begin and even if there is plentiful seed fall, it may not survive the ravages of scavengers such as mice and grey squirrels. The ground needs to be prepared by light cultivation and you will need to control competing weeds. Perhaps the best advice is simply to take advantage of any regeneration that does appear rather than trying to make it happen. Even experienced foresters can fail to regenerate stands by natural means!

WHAT MAY NEED DOING AS A STAND OF TREES DEVELOPS

Here we look at four silvicultural operations that a stand of trees may require at some stage. They are not equally important. For example, high pruning may only be needed for reasons of rideside safety or if you hope to grow high quality timber.

Cleaning

Cleaning makes us think of washing clothes, tidying our homes, scrubbing a work-surface, or polishing the car, but in forestry parlance it has a special application. Cleaning is the job of removing unwanted growth in a stand of young trees that's past the establishment phase but before trees are big enough to be usable, when approx. 3-7m tall. It is an operation that should follow weeding but is commonly neglected although much needed in young woodland. It is ideal autumn and winter work and can be done reasonably safely with hand tools such as saws, hand axes and billhooks, secateurs or pruners. Anything from a tiny group of perhaps a dozen trees upwards will benefit from cleaning.

Cleaning is all about sorting out a stand of young trees to help those you want to succeed at a time when those not wanted are small enough to be dealt with by hand.

I thought the best way to illustrate it is to reproduce an extract from Chapter 15 of my book *What Happened to Our Wood*. The chapter is called 'Cleaning Taid's Wood'. I've added some headings for clarity.

The story so far
In 1987 my brother-in-law and I planted 4ac of new woodland with a mixture of ash, wild cherry and oak trees. We called it Taid's Wood after my father. The trees grew vigorously in their early years with each individually protected by the then latest device, the plastic tubes or tree shelters that today litter so much of our countryside. The tree shelters worked well, hardly any trees died and many grew nearly 1m in height each year.

Sporadic removal of tree shelters began in the mid-1990s as the cherry trees quickly swelled to fill the tubes. Removal of all of them from Taid's Wood began in January 1999. But, once de-sheltering was complete, we rather neglected this part of our wood.

The problems neglect of woodland can bring
Neglect can be damaging for broadleaved stands like ours. Unlike most conifers, species like oak, beech, and ash often fail to develop a straight stem unless deliberately helped by the way the trees are grown and cared for. Straight stems are encouraged by growing trees close together, to force upward height and suppress heavy side branches by mutual shading, by selecting trees of good genetic stock, or by formative pruning. In Taid's Wood although we had planted the trees about 3m apart, coppice growth of sycamore and some hazel did in places provide the desired more densely stocked conditions. But using other woody growth runs the risk that it will dominate, and even suppress, the planted trees. It's like allowing moss to green up bare patches of lawn only to find it takes over everywhere. Thus a time comes when a stand of trees needs sorting out, and the job is called 'cleaning'. It brings in no cash but lays the foundation for the future crop, and is a job neglected at peril. In Taid's Wood serious cleaning began in the winter of 2000. The best ash and

cherry were already 10m tall and the oak about 7m.

What the cleaning entailed

Initially, one is hesitant in deciding what to cut but over several half days work, six distinct activities evolved. It was like tidying the garage, the more one does the more is discovered that needs attention and the whole job takes longer than intended. The main task in Taid's Wood was to cut regrowth of sycamore and hazel and self-sown birch and sallow that competed with the planted ash, oak and cherry trees. My small Stihl chainsaw, lightweight with a short 12in guide bar, proved ideal. Stem thickness rarely exceeded 3in and with a well sharpened chain the saw cut through in seconds. With head down and felling several coppice stems in quick succession a rhythm develops though at the risk of being overzealous. Not only is there the question of safety, but the need to consider carefully each stem to be cut, since not every unplanted woody stem is necessarily unwanted. With the fairly widely spaced trees there was still benefit to be had from extra stems to make good any gaps and, in a few places, to substitute the planted tree if it was poor. It was important continually to keep an eye on the surrounding trees where one worked and always on the look out to recruit a new stem. Several well-grown sycamores were added in this way. The grand object was to secure the future of all the best trees and sometimes bring into the reckoning a new one. By cleaning, an unmanageable thicket was turned into a young stand of forest trees.

A few of the planted trees, although growing vigorously, had such badly shaped stems or ungainly crowns of branches that no amount of formative pruning would produce anything worthwhile. These 'wolf' trees are best cut out during cleaning. If not, they occupy much space and interfere with better formed adjacent trees. Perhaps one ash in 20 was culled – 'de-wolfed' – from Taid's Wood and one in 50 of the generally better cherry trees. None of the oaks was really big enough to merit the de-wolfing epithet and also they were being grown in a different way from the rest of stand.

Tiny groups of oaks had been scattered throughout the matrix of ash and cherry trees at 14m centres with the intention that oak will form the final crop. At each centre three oaks were planted in a tight triangle with the idea of selecting the best. A choice of one out of three for oak, with its generally poor form, is really too few. I hoped to get away with it because the acorns had come from good genetic stock and the activities of cleaning and formative pruning at the right time had been factored in. Assuming I was around, Taid's Wood for one would not be neglected! As cleaning proceeded, the triplets of oak were inspected and one or even two of the trees cut out if there was already a clear winner. However, for a few groups it was too late and all three oaks, with their slower growth, were dead or nearly so, suppressed by vigorous adjacent ash or sycamore.

All the while cleaning proceeded, the job was repeatedly interrupted by the strings and ropes of *Clematis* that entangled the undergrowth and ascended into the crowns. Sometimes they stretched from tree to tree like rigging on a sailing ship, but without the order and purpose, and would prevent cut stems falling to the ground. The problem was not serious since during establishment of Taid's Wood killing *Clematis* had been a priority. Nevertheless, dealing with this temperate vine was a crucial part of the cleaning. Tough stems, an inch across, were cut near the base and, importantly, the climber itself was pulled down from every tree it had assaulted. On one wintertime visit it was almost fun. Ben, our youngest son, had come to help and took on the climber-pulling task. It was the Saturday before New Year when most of the country was still in the grip of snow and ice. The trees in Taid's Wood were not only gripped by *Clematis* but still, albeit loosely, by Wednesday night's snow. Cutting any woody growth with the chainsaw sprinkled snow but Ben, yanking and pulling down *Clematis* repeatedly got showered! At least the snow did lubricate the branches so that the whole vegetable mass slid off the tree quite readily, and the hard work kept us warm.

The final cleaning task was to revisit all the better trees. While several ungainly 'wolf' trees were felled, many ash and a few of the cherry were forked or had heavy branches that interfered with less vigorous but better formed neighbours. Although the offending limbs were often rather large – ideally pruning should be confined to branches less than 2in across – removing them gave room for adjacent trees to develop. When the first proper thinning takes place in a few years time the future of these partially de-limbed trees will be re-considered, and many will probably go. As with thinning, the intensity of cleaning and what is removed is an open question. Very thorough cleaning and removal of all the poor trees, could make a lightly stocked stand decidedly gappy and moth-eaten. As with most forestry tasks, today's handiwork will be judged many years into the future.

Pruning

In woodland operations there are three situations when pruning might be undertaken, none is essential compared with pruning roses, grape vines or fruit trees.

- While cleaning a young stand of trees, singling a fork in the main stem of an otherwise desirable tree is a pruning that can be done with secateurs or long arm pruner.
- For reasons of safety and good access, pruning branches from ride and track side trees is commonly needed.
- When growing high quality timber, side branches are pruned from the lower trunk to a height of about 4-6m to restrict the knotty core of timber. This is done early in the life of the stand while the trunk is no more than 10-15cm thick.

When to prune
In terms of season, pruning of native trees can be done at any time though it is best to avoid late spring when copious sap flow can occur. For walnut and wild cherry trees pruning must be done in June or July to minimise disease risk.

In terms of size of tree, this will depend on the purpose listed in the bullet points above. One rule-of-thumb is that it is best not to prune branches bigger than 5cm at the base, otherwise they take years to heal over and readily become infected and start to decay.

How to prune
The aim is to sever the branch cleanly, close to the stem. Do not do this absolutely flush but angle the cut slightly outwards beginning just outside the branch bark ridge – the zone of creased bark at the junction of branch and trunk.

Special long-handled pruning saws do the best job since you want to cut through, not tear, the bark. Lopping with an axe is dangerous and generally damaging.

The pruning saw I use, it's a superb tool.

Use of sealants
Although painting a sealant over a pruning scar feels as if you are providing a protective bandage, research has shown that little is gained, either in preventing disease or hastening the callus that heals over the wound. Spend time on doing the job carefully and cleanly, not money on proprietary sealants.

Thinning

This topic and the next one, felling, are touched on only briefly. Both are major operations that require professional advice, trained operators and are usually done by contractors. What I will do is explain the idea and point you, I hope, in the right direction.

What is thinning?
At planting, or from natural regeneration, commonly 2-3,000 trees are established per hectare. Yet at the end of a tree's life – at its rotation age – 50 or 100 or more years later when it is big and ready for the market, only 50 to 250 trees per hectare are left. The process of reducing the stocking of trees down to this small number is the operation of thinning. It is done at intervals as the stand of trees gets older and older.

If you are wondering why one starts out with so many trees when only about 10% are wanted at the end, there are three main reasons: To occupy the available ground quickly; to provide a wide choice of trees from which to choose the very best, and to prevent individual trees developing thick, heavy branches and a spreading crown which happens if, in early life, they are spaced far apart without competition from neighbouring trees.

How do you thin a stand of trees?
The grand aim is to remove the worst and leave the best. You cut out dead, dying and diseased trees and those of bad shape. Also you try to favour the best trees by opening up around them to give their crowns space to grow.

If you open up a stand too quickly by thinning very heavily, the remaining trees – the ones you are wanting to favour – will be blown about, snap off and sometimes be uprooted. This is exactly what happened in a stand of ash, oak and sycamore in my wood that was thinned in the very wet and windy winter of 2013/14. The contractor had almost finished the job, then the storms came and another 30 or so trees blew down which, fortunately, he was able to clear up and use, but the stand now is too open and gappy. In general it is usual only to thin out (remove) about one-third of the trees in a stand at each thinning.

When do you start thinning and how often is it repeated?
Thinning begins when trees are 10-12m tall and are big enough to sell for firewood or pulpwood. To begin with the operation is repeated every 3-5 years but as the trees become large and nearer to maturity the frequency of thinning is reduced to every eight or even 10 years.

Does thinning earn money?
Yes, usually, but not a lot. Because trees are being cut out from among others left standing, the operation is quite tricky and the total amount of wood obtained per hectare may be only be 20-40 tons, thus merchants or contractors won't pay a lot for thinnings. Worse, when trees are small in the first or second thinning you will be lucky to get any income at all since the cost of doing it may be more than the trees are worth. However, by thinning you are improving the crop for the future, so it is best to do the job when you can and not wait in the hope of a better price round the corner.

However, the rapid rise in demand for firewood in the last few years has increased timber prices very considerably and once unpromising woods can now be thinned profitably. As we comment elsewhere, this is a blessing for conservation since dark, neglected woods are less favourable to biodiversity than well-thinned ones with light reaching the woodland floor, plenty of glades, and open tracks and rides.

Felling and starting again

In commercial forestry trees are felled at the end of the rotation. Usually in Britain all the trees in a stand are felled at the same time: this is called 'clear felling'. It is a drastic change to your woodland but is the most profitable way if this is important. After felling the ground is replanted unless natural regeneration is present or can be expected.

Because clear felling is a sudden and total opening up, there is a trend towards less severe forms of intervention. Most are covered by a new term 'continuous cover forestry'. As the name implies, trees are felled and regeneration encouraged without completely clearing the stand: some big trees are always left and a woodland appearance is maintained. It is a more intensive system, requires skill, and may be less profitable. One way is to fell small groups of trees and replant in the middle of the openings, another way is to thin heavily among the old trees and develop volunteer regeneration, whilst a third way might be to clear fell most of a stand but leave sufficient old trees to give the appearance of continuous cover, and so on. Which system will work best depends mainly on the site and the tree species involved. Space does not allow more details here. To illustrate the complexity, one can't regenerate woodland using small openings if the species concerned is ash or oak, both of which require plenty of sunlight from an early age, but it could work well with the more shade tolerant beech. That said, trying to establish groups of tress in small gaps is usually doomed to fail. A useful rule of thumb is that the minimum diameter for a gap to be worth planting (regenerating) is 1½ times the height of adjacent trees.

If you think your trees are mature and your wood has reached the time when it should be regenerated – or you simply want to start again and leave your own mark on the wood – seek professional advice. As I have said before, an hour of a professional forester's time will repay handsomely as alternatives are presented and opportunities discussed. Will and I would say this as we are both chartered foresters(!), but we believe it to be absolutely true.

COPPICING AND POLLARDING

Coppicing

These terms describe cutting a tree near ground level (coppicing) or someway above animal grazing height (pollarding) – and so out of reach of browsing animals, especially cattle – in order to use the shoots that emerge to form the next crop. The words coppice and pollard are also nouns that name a crop or tree created in this way. Only broadleaved trees can be coppiced with the famous exception of redwoods, monkey puzzles, some cypresses, Chinese fir and one or two tropical pines. Even with broadleaves, coppicing potential and vigour varies. At every coppicing, a few stumps die, and species such as beech and common alder generally do not sprout new growth very well. With beech it seems to coppice better in the wetter west parts of the country. Certainly in my wood in Hampshire, the southeast, I have been surprised how many stumps of beech have coppiced following wintertime thinnings in 1997, 2003 and again in 2011.

Restoring an old coppice
I begin with this because if you have bought a wood in the lowlands it is quite likely that it was once managed as a coppice. This is also true for many woodlands of native broadleaves in the uplands of southwest England, Wales, the Lake District and western Scotland. Indeed, historians believe well over half of all woodlands were managed by coppicing in medieval times. Yet today, active coppice management is confined to about 25,000 ha or only one per cent of the total forest area.

A rapid decline in coppicing occurred in the early decades of the last century. This arose mainly from a collapse in markets for tanbark, charcoal, firewood, and rustic products such as thatching spars, hurdles, and besoms. There is now something of a revival of these rural crafts and coppicing is back in fashion along with all its wildlife benefits.

Why restore coppice?
There are several answers to this question. First, coppicing will bring back a neglected rural craft and demand for rustic produce such as hazel spars and hurdles for fencing, whilst poles and sticks

for the vegetable garden are on the rise. Secondly, a coppice is the perfect renewable and sustainable energy source for firewood or charcoal – see Will's chapter (8) for an excellent account. Thirdly, and most importantly for many people, restoration of coppice will promote a wildlife habitat, a theme we return to in Chapter 9.

Can a wood be too small to consider for resuming coppicing? In a word, no. If there are clear signs that a wood was once coppiced, even a 0.2ha copse (as the name implies) can be restored again, perhaps as two 0.1ha coupes cut five years apart. If coppiced once, it can be coppiced again.

Recognising coppice woodland

Because coppice uses the regrowth that sprouts from cut stumps, the key to identifying that a woodland is in fact a coppice is to look at the trees and how they appear. In a forest that is *not* coppiced but felled and replanted or regenerated by natural seeding, all the trees will have single stems or trunks at ground level. In a coppice things will be quite different. Most trees will be multi-stemmed at or near the base with two of more trunks growing up very close to each other.

Sometimes in a long neglected coppice one tree stem or trunk will dominate and the others from the stump fail to grow much. Often the dominant trunk will appear slightly bent or swept at the base curving up a little from the ground and slightly off-centre. This feature, and the presence of even weak basal shoots, indicates a likely history of coppicing. A good example is the National Trust's lovely Hembury Woods on Dartmoor. They have been well thinned over the last 40 years with the best trunks of coppice origin favoured (a treatment called 'storing coppice') and the woods now look like a delightful high forest of oaks. But careful inspection of the base of each tree shows this off-centre feature of the trunk and often several straggly shoots persisting and growing out of the base – the sure signs of coppice origin.

Where coppicing has been practised for hundreds of years the original stumps will have long disappeared. However, the present trees may still be in multi-stemmed groups and may also make an informal ring with other clumps a bit like a fairy ring of mushrooms and toadstools. What has happened is that at each coppicing the new shoots arise at the periphery of a stump and so it enlarges a little each time.

Types of coppice
Coppices are not all the same. They vary in species composition, in their rotation or length of cutting cycle, and whether or not they have large trees (standards) scattered through them. I give a simple classification below that I hope is helpful. Once you know the type of coppice woodland you have, then its treatment becomes clearer.

> *Pure coppice* – This is a coppice made up of one species. Examples include sweet chestnut, which is common in Essex, Kent and Sussex, hornbeam on the heavy clay soils around London, hazel in many woodlands in the southern half of Britain, and oak woods all along the western seaboard.

> *Mixed coppice* – This is a coppice of several species and deliberately managed as such. It is fairly common and is usually a short rotation coppice for producing bean poles, pea sticks, hedging stakes and similar small sized material. Mixed coppices typically contain hazel, birch, sallow, hawthorn, ash and sometimes oak, lime, field maple, buckthorn, wild cherry and alder.

> *Short rotation coppice* – While there is no strict definition of what 'short' is, it is usually accepted that coppice crops cut on a cycle or rotation of less than 10 years are short rotation, typically hazel. However, the name has been hijacked by the energy industry as 'short rotation coppice', or SRC as it is sometimes called, and is recognised today as a special crop of poplar or willow grown for biomass or 'green' energy.
> To be strictly accurate, we should mention osiers, where special varieties of willow are cut each year to yield the pliable and wiry 'withies' for basket making. This is still an active local industry on the Somerset Levels.

> *Long rotation coppice* – All other coppices are worked on cycles longer than 10 years, and sometimes as long as 30 years, but rarely more. Such coppices include sweet chestnut, cut on a

rotation of 12-18 years depending on quality, and worked for fencing stakes, building and traditionally hop poles, oak on a 25-30 year cycle when it was managed for tanbark, and hornbeam a similar length of time when it supplied firewood and charcoal for industry and the hearths of the home counties. Other coppices managed for firewood, or even pulpwood, would have rotations of 20-30 years to produce pole size material.

Coppice with standards – This is a coppice crop within which there is a scattering of larger trees that stand over the coppice or understorey (called 'underwood' in such circumstances). These larger trees, called standards, are well spaced apart so that the coppice is not heavily shaded. The idea is that both large timber as well as small-sized produce from the coppice are obtained from the site. Coppice-with-standards make an attractive stand and many traditional coppice woodlands were managed in this way.

It is easy to tell whether a woodland today is a neglected coppice-with-standards. As well as having clumps of multi-stemmed trees arising from the base, the big trees present will, very tellingly, have large (usually dead) branches low on the trunk. Typically these large limbs will be only 6 or 7m from the ground. This shows that once these trees grew in much more open well lit conditions and not in a dark forest stand environment, even though today they may be in dense woodland. Part of my wood that we call Nain's Copse is like this with an underwood of sycamore and ash, last cut in about 1960, and a scattering of oak standards with only short branch-free trunks.

Standards are carefully managed along with the main coppice crop. Each time the coppice is cut, say every 20 years, the oldest standards will be taken and a few new young ones recruited. Typically a coppice-with-standards will have several age classes of standards that match the cutting cycle. Thus a coppice cut on a 20 year cycle should have a few standards each of 20, 40, 60, 80 and perhaps 100 years old. This is illustrated at the end of Chapter 2 and again here to try and make it clear.

Oak is much the commonest species used as a standard.

Ash is also suitable, but beech and hornbeam cast too much heavy shade.

Coppice with standards showing different ages of standards.

How to coppice

This is a wintertime job though it can be carried out later into the spring than planting. Ideally coppicing should finished by the end of April otherwise the new shoots emerge late in the summer and are still small by the following winter. But most importantly, if the tree is coppiced too late into the spring, the sap will be rising and the stump may bleed, especially birch. Cut coppice stems cleanly to near ground level taking care not to damage the bark. The one exception is hornbeam when a 30-50cm (12-18in) stump should be left. Provided the area is not heavily browsed, new shoots (coppice) emerge from the stump, called a 'stool', in late spring. The new shoots may grow 1-1.5m (4-6ft) in the first year. Initially many shoots will grow up but these thin themselves often to leave half-a-dozen by about 15 years of age or two or three times this number in the case of hazel.

Standards can be either recruited by leaving the occasional coppice stem or be planted.

Do be aware that new coppice shoots are very popular food for a number of woodland animals (as well as trespassing sheep). You need to make sure that new shoots are protected as they grow in just the same way as newly planted trees need protection until they have grown out of harms way. You can do this by 'dead hedging' where any coppice material, tops and other waste you haven't removed from the site is piled over the stool as a deterrent. It works well for a small number of stools, but for larger coppices recourse to fencing is likely to be necessary (pages 56-57).

Pollarding

When a tree is pollarded, it is 'coppiced' above animal grazing height. This takes place at regular intervals of 10-20 years and the shoots (branches) are traditionally used for firewood or stored for winter cattle feed. The operation fell into neglect and many of today's large stag-headed, multi-limbed trees are the result. Let's remind ourselves what pollards look like from John White's sketches in Chapter 2.

Different kinds of pollards

Pollarding is seeing a revival, particularly for riverside willows and there are some attempts to resume the practice on long neglected trees, but is it not at all easy. The trick is to cut off the branch just above the old cut and not below it into even older wood, but even this will not guarantee success.

Pollarding can create very old and squat trunks, often decay ridden but still alive and a serviceable shell with an interior home to highly specialised beetles and other micro-organisms. It is a good operation to perpetuate.

Speckled wood butterflies are common in most woodland especially where there's plenty of dappled shade, tracks and rides.

7

Woodland Crafts and Produce

This chapter is all about what you can obtain from your wood. We shall first look at topics relating to timber production that we mentioned in Chapter 3, namely:

- How to assess quality and quantity of timber;
- Advertising trees for sale in magazines like *Woodlots* and via websites;
- How to sell trees to contractors; and
- Typical prices for timber.

We then look at the rewards of using you own timber whenever you can, commenting on woodturning, making and using stakes, and finally using green timber. This eclectic group arises from personal interest and by way of illustrating potential since we can't cover everything in a short book. We won't say much more about firewood, it is now so important that Will has written the next chapter (8) just on this topic. Also we won't say anymore about Christmas trees that we covered in Chapter 3.

The basics of timber production

I have mentioned more than once that inviting a professional forester to cast an eye over your wood, even for only an hour, will handsomely repay itself. You will have a good idea if there is timber of worth, when and how it is best managed, and what steps to take, and you'll probably learn lots more besides. But even if timber production and earning income is of no concern, knowing what you have or don't have, as the case may be, enriches your knowledge. What I do here is provide rough and ready guidance to help get you in the right 'ball park' as the Americans would say.

If your wood is newly planted or none of the trees is over

8m (25ft) tall you can skip the next section since your trees are still too small to be saleable as timber. The exception is hazel coppice, but we will comment on this and other coppices a little later.

How much timber do you have?

In the trade timber is mostly sold by the cubic metre of volume that, for freshly felled unseasoned logs, equates very, very approximately to a ton in weight. We will stick with this more familiar measure. What you need to do is work out how much you have of each main species. You need to know the number of trees and their average size.

To find out the number of trees in a small wood, and if trees are in rows, you can count every one or (say) every fifth one and work out the total. If rows are not obvious sample your wood by randomly[1] locating square plots of 10 x 10m. Mark the corners and count the trees inside the plot. For a 2ha (5ac) wood lay out 10 plots, for a 5ha (12ac) wood 15 plots should do unless the wood is very variable. Now calculate the average number of trees per plot, multiply by 100 to discover the number of trees per hectare, and then multiply by your wood's total area for the total number of trees.

Once you know the number of trees of each species, you need to estimate the average volume or weight of timber. Here we become very imprecise, but you'll at least get an idea. When counting every fifth tree or the ones in a plot, measure their diameters as well – diameter is measured 1.3m above ground. You can use an ordinary linen tape measure and your GCSE maths to convert circumference to diameter! The table below gives an approximate weight of timber according to diameter of the trunk. I originally drew up this table for an article about estimating amounts of unseasoned firewood *if all parts of the tree are used*, but it will hold crudely for timber in general.

---

1 An objective though not entirely random way is to begin at a corner of a wood. Then take as many paces into it equal to the first two digits of your credit card. Where you stop is your first plot. Pace on the distance equal to the next two digits on the card for the next plot and so on. If you reach the other side of the wood, turn round, take a different direction, and continue until you have 10 or more plots.

Diameter of tree (cm)	Approximate weight of wood (tons)
10	0.02–0.04
20	0.2–0.3
30	0.6–0.8
40	1–1.4
50	1.8–2.4
60	3–4
70	5–7

The table shows that trees with diameters greater than 20cm are pretty big and ones above 40cm very sizable indeed. Unskilled operators should not attempt to fell trees any bigger than 10cm diameter at the most.

To complete your rough estimate, I suggest you total up how many trees are around the 10cm diameter mark, then how many are around 20cm, and so on and then multiply the number in each diameter class by the weight given above. This will give a better result than working out an overall average diameter to get an average weight.

Remember, all this is very rough and ready, but it will tell you whether you have something like 100 tons of timber in your wood rather than 500 tons, or vice versa. The illustration shows piles and piles of logs at the entrance to my wood. At the time I only owned 22ac of woodland, and about 630 tons of logs came from just part of it and from just removing the pines from among beech trees, it was not a clear felling. Although many readers will not be interested in this side of forestry, eventually all woods need some thinning out and a lot of timber can result.

Will rightly adds at this point that if you want to go deeper into this subject looking at how to estimate timber volumes, weights and similar data there is no better source than the Forestry Commission's *Forest Mensuration – A Guidebook for Practitioners* by Robert Matthews and Ewan Mackie. It is not a light read and will have much material you probably don't need.

*Not a priority for many, but using or selling your own timber
can help with income and reduce Britain's imports.*

What makes timber good quality for selling?

I added 'for selling' to the subheading because for many purposes,
such as wildlife, amenity, or firewood, timber quality is irrelevant.
But if you want to sell timber, and many small woods contain
trees of value, quality is the key to price. It has four elements.

Quality firstly means species. Of the broadleaves, good
oak and ash fetch better prices than good beech or birch, while
willow – apart from the specialised cricket bat market – is
virtually unsalable except as biomass. Of the conifers, Douglas
fir and larch usually sell readily, pines can be more difficult, and
while good spruce is in demand, poor material is only fit for
chipping. Added to this are specialised markets for prized timbers
like walnut, sycamore with a curly grain, wild cherry with vein,
but don't worry about these unless it interests you.

Secondly, good quality means large size. Small trees are of
little or no value regardless of species. Prices increase rapidly with
size. An oak of 25cm diameter may sell for firewood while a good
one of 75cm may go for several hundred pounds. Occasionally a
tree can be too big, like a large old hulk of a seaside pine, simply
because sawmills cannot cope with the dimensions.

Thirdly, good quality means defect-free. A log must be free
of decay or bad deformity and big knots.

Finally, good quality means straight. Straight logs are
transported more efficiently and when sawn the outturn from
them (proportion of usable timber) is much higher than from
bent or forked ones that result in a lot of waste.

So, a buyer looking at your trees will be keeping an eye out
for large, straight, blemish free trees of their species of interest.

How to advertise trees for sale

Most timber sold by private owners, is by 'standing sale'. That is the owner sells the trees while they are still standing. The job of cutting them, extracting the logs, and hauling them to a mill is done by the purchaser. It saves the hassle of doing the dangerous operations yourself. Of course, big companies and the Forestry Commission have their own felling gangs, but I don't expect many readers will be in this league.

The minimum a buyer needs to know is which trees are on offer. So you need to identify these and mark them clearly, say, with a spot of paint on both sides of the trunk. Everything else about your wood, whether you have the appropriate felling licence, the quality of access, the terrain, the total amount on offer, will be evident when the potential buyer visits, though it helps to have this information in advance. But how does one find a buyer in the first place?

Never underestimate your neighbours, so ask around if someone has sold timber in the last couple of years. If it went well, you have your recommendation. Failing that some Forestry Commission offices maintain lists of who might purchase timber. If a professional forester has visited your wood, do raise this topic. Trade magazines such as *Forestry Journal* and *Forestry and Timber News* have 'Timber wanted' adverts as do *Smallwoods* and the *Quarterly Journal of Forestry*. Following up an advert will quickly tell you what is being sought, and so set you on the right track.

For my part in the southeast of England the *Woodlots* directory run by Woodnet, a grouping of growers and wood users, is excellent.[2] It's like the *Exchange and Mart* of forestry and is particularly helpful for the small owner. Other regions have networks or local associations that help timber sales like *Yorwoods* in Yorkshire, Cumbria Woodlands, Chiltern Woodlands Project, Sylvanus Trust in the West County, and Anglia Woodnet. Regrettably the very successful '*Ecolots*', which grew out of the '*Woodlots*' idea, ceased in 2013. Until I began selling to a local contractor who I knew well, every advert I'd placed in *Woodlots* brought a sale even if not the price my inveterate optimism hoped for!

Will adds that it's well worth mentioning the Sylva

---

2 www.woodnet.org.uk

foundation's 'myforest' tool[3] that has directories of contractors, timber sales and timber wanted as well as a useful online woodland management tool.

Many large private owners contract out woodland management, including selling timber, to a consultant or a management company such as UPM Tilhill, Fountain Forestry or Scottish Woodlands to name but three. For small woodland owners not only will this be costly, but it takes away part of the very fun that having a wood brings.

Selling to contractors
One of the easiest ways to get in touch with contractors is via adverts in *Woodlots*. Most are highly reputable and will be willing to visit your wood provided you can offer for sale at least 20 tons of decent timber. Of course they prefer 200 or 2000 tons, but a good lorry load is the minimum most will consider. After looking at what you've got and assessing factors like access, if they are interested they will offer a price of around a £5–30 per ton, unless what you have is exceptional.

Once you accept an offer, a contract needs to be drawn up making explicit: what is being sold, the felling licence approval, how long the purchaser has got to take his trees (two years is normal), when payment is due, what insurance cover the purchaser must have to indemnify you against accidents, how tidy the contractor should leave the site, and other terms and conditions. A map showing the area where the trees are and the access routes is essential. These maps are usually available from your local Forestry Commission (or Natural Resources Wales) office to support felling licence and grant scheme applications.

Instead of advertising you can circulate details directly to local contractors. These are rather like an estate agent's particulars for a house sale, and describe what is on offer and what special conditions may be laid down. For example in my wood the main track must always be kept free for Network Rail staff, so timber operations must work around this.

Typical prices for timber
I'm not going to say much, other than give you an idea of prices of standing trees. Remember in <u>very general</u> terms a

---

3 www.sylva.org.uk/myforest

'ton' equals 'cubic metre'.

Small sized material for firewood	£5–30 per ton
Medium sized trees for fencing, chipping and small logs	£10–20 per ton
Large conifers for sawmilling	£15–60 per ton
Large hardwoods for sawmilling	£30–100 per ton
Large hardwoods of exceptional quality	up to £250 per ton

Assessing coppice
Coppice is sold by area, around 1ac (0.5–1.0ha). A buyer will look at it for density of stems, their straightness, freedom from kinks (especially hazel), thickness and such factors as ground conditions, proximity to a track and so on. However, there is an enormous amount of neglected and overgrown coppice unfit for traditional usage which needs bringing back into rotation first before it is saleable for anything other than perhaps firewood which, as Will outlines in the next chapter, can be a lucrative market.

Using your own produce

Owners of small woodlands, possibly more than those of large estates, have a yen to use their woodland's produce as well enjoying the woodland itself. This desire can lead to interesting results and more than a few benefits. Of course, the possibilities depend on the type of woodland, but let's look at some of them.

Supplies for the garden and home
If you have hazel coppice, or any type of coppice or young woody growth, finding bean poles, pea sticks and other vegetable patch and gardening props and supports will be easy. But what about rustic fencing, or your own decking, or making a seat out of a fallen oak – the sapwood might decay in a few years but the heartwood will remain for at least two decades and often longer. Talking of durability, sweet chestnut is the equal of oak in this respect and Lawson cypress may even better it. So there is a use

for the poles from the overgrown hedges of suburbia!

These remarks about fencing were prompted by using Douglas fir from my own wood some years ago to replace the fence at its entrance once a beech thinning was completed and the last lorry had uplifted the final consignment. Douglas fir is moderately durable and I hope will last about 15 years. The trees used were 47 years old and were ones in the subdominant or suppressed crown class[4] but still alive, with long slender stems to make rails, while thick enough at the base to produce a post or two. Despite their slow growth from a life in heavy shade these otherwise valueless trees had developed good heartwood. If the job had been in early winter the tops of one or two of them would have made a passable Christmas tree. But I did the fencing in April and May and just contented myself with relishing the resin-rich aroma of Douglas foliage.

And the thought of foliage leads naturally to decorative uses of it in the home along with bark, cones, mistletoe at Christmas and other gatherings from the wild.

Energy

Perhaps collecting firewood is the most obvious use for your own produce, always provided you have a house with a fireplace and chimney! It is important to remember that firewood is a green form of energy. If you cook or keep warm by burning wood, you are not using a fossil fuel like oil, coal or gas, so your net impact on carbon dioxide increase in the atmosphere is neutral – a tree grows by absorbing carbon dioxide through its leaves to make wood. Will covers this thoroughly in the next chapter.

More adventurous

So far I have confined remarks to produce that you can easily obtain without recourse to contractors or taking a chainsaw course. But clearly there is no limit in reality: friends of ours bought a dilapidated house with a 23ac coppice with standards just weeks before the great storm of October 1987. The oaks felled by the storm provided all the flooring needed in the restoration of the property.

---

4   These are trees that have lost the competition for light in a woodland stand and tend to be slender with small branches and growing in the shade of adjacent larger trees (Will Rolls).

Wood for turning

There is a special delight in a gift of a bowl or goblet or paper knife or delicately crafted ornament carved or turned from wood from one's own patch. My brother received an ash 'thing bowl', as we called it in our childhood, for his 60th birthday. This is the most recent of many gifts to relatives and friends turned from pieces of yew, oak, wild cherry and even birch from the wood. There are woodturners galore. So popular is the hobby that at least three dedicated magazines can be found in a large newsagents and as a quick 'Google' search will reveal.

Most woodturners acquire their material as off-cuts from sawmills, DIY stores, fallen trees, and other bits and pieces. Garages become cluttered, and garden sheds inaccessible, as all sorts and sizes of wood are stored waiting their turn and, importantly, slowly seasoning. My brother-in-law's late father-in-law was a keen woodturner and he used old toilet seats as a wonderful source of tropical hardwood! But as a woodland owner, can one sell material to this particular market?

What wood turners look for

Most woodturning seeks to display grain, figure, colour or patina, or an unusual feature like a knot cluster. The woodturner's expertise is seeing the potential and, like a diamond cutter, using his craft and skill to bring it out for all to admire. But, few woodturners give much thought to where their raw material originally came from and make the link back to the tree itself. They appear content just with the piece of wood that holds interest.

I am fortunate that twice the Hampshire Woodturners Association have visited my wood. I want to share this in a little detail to show what a delicatessen woodland can be. Within reason I offered to fell any tree, standing or fallen, if it might hold some turnery interest. I was armed with a chainsaw. This is what happened.

1. The first tree was a small beech about 15cm in diameter that had died about two years before from beech bark disease. I cut it at the base and whoops of delight revealed that the feature known as 'spalting' – dark lines like an artist's fine pen and ink sketch – were present. Several 2ft lengths were

cut off the stem for the woodturning visitors. The spalting feature occurs at a very early stage of 'decay', so does not continue when turned into an artefact or brought indoors.

2. The next tree was an oak with a burr – a nobbly swelling on the side of the trunk – that I thought they would like. I was right in thinking that burrs are interesting, but this one was too small and no one wanted it. Large burrs that swell out from a tree like a heavily pregnant mum, are much sought after.

3. Farther into the wood we passed an uprooted birch, perhaps 40 or 50 years old, that was leaning steeply but not on the ground. I was asked to cross-cut it. This was dangerous, as is dealing with all hung up or blown trees. But this birch had been prostrate for about 18 months and so, gingerly, I began cutting into the trunk. Eagle eyed I watched for any shifting of weight or tightening of cut, but there was none. The passage of time must have eased the huge tensions uprooted trees possess, and I safely cut right through. The cross-section exposed a jigsaw of pale yellow, pink and ochre patches across the surface! Several sections were cut and added to other trophies destined for the lathe.

4. Nearby was a 70-year-old ash with a huge limb rising from near its base. It was of coppice origin and I remarked that the tree would be much improved if this limb was removed. So remove it I did and was asked to cut it into 4-foot lengths. Subsequently the lengths were split into sixths, using a sharp wedge and mallet to provide cleft ash for turning into chair legs.

5. The next tree was a long fallen yew that had blown down in the storm of 1987. Much was still sound and the wood, with its deep pink heart, is always of interest. Several pieces were cut from branches.

6. The last stop was where a great oak had been felled five years before, the wood of which possessed the rare and valuable feature called 'brown oak'. This is a natural chocolate brown staining caused by the beefsteak fungus. Discarded wedges from the felling still lay about – at least until my woodturning visitors found them!

Lessons for a woodland owner

I learnt much from the visits of the woodturners. They look at trees and bits of wood with a different eye. It has added to my own interest in our woodland, and not least that I could make a little money from trees that were totally unsaleable to the usual timber markets. But before one gets too excited, woodturners are quickly satisfied in terms of quantity and the size of this niche market for any one woodland is unlikely to be large. Nevertheless, I hope that relating the woodturners visits in the way I have has shown you that there are other products that can come from a small woodland or even the odd tree. The usual premium placed on large, defect-free, straight stems has no place, quite the opposite in fact.

Making and using stakes

Every woodland owner can make use of stakes. They are needed for fencing, for corner posts or gateposts, for name boards or fixing warning signs, while short stakes are used to stop vehicles encroaching soft ground. Thin stakes are needed to support tree shelters and other guards of newly planted trees. And, that's not all: birds perch on them, and where else do you put a mug of piping hot coffee? Most if not all the stakes you'll need can be obtained from your own wood.

What makes a good stake?

A stake, like any other product, must be 'fit for purpose'. This means it must be: straight and of the right dimensions; strong, and; durable i.e. resist decay, even in ground contact, for many years. Other desirable qualities are an ability to hold nails, be resistant to splitting, and for the surface to be free of splinters, snags and other hazards.

This yellowhammer knows what a post is for.

Tree species that make good fencing material
Most conifers (softwoods) produce straight uniform stems, therefore, make potentially good stakes. Not all are naturally durable, and unfortunately not all that lack durability will accept preservative well. For broadleaves (hardwoods) the situation is even more variable, and it is also more difficult to find and cut a high proportion of straight stems. Below are some species worth considering. Ones left out like spruce, birch and poplar are unsuitable.

Species	Straightness	Natural durability*	Uptake of preservatives	Other factors
Conifers				
Pines	very good	fairly poor	fairly resistant	
Larches	very good	good	fairly resistant	good stakes
Douglas fir	very good	fairly good	resistant	
Cypress**	good	good	fairly resistant	good stakes
Yew	moderate	good	resistant	fine gate posts
Broadleaves				
Ash	good	poor	fairly good	may split
Beech	moderate	poor	good	
Oak	poor	very good	resistant	use heart-wood fine gate posts
Sweet chestnut	good	very good	resistant	good stakes
Sycamore	good	poor	good	

*   natural durability: poor = decays in under five years; good = will last about 15 years

** Lawson and Leyland cypress and the similar red cedar make durable stakes

Sizes of stakes and other fencing materials

Size depends on usage: for example gate posts, strainers or corner posts will need to be thicker than fence posts simply acting as wire supports in a long run of fence. The following sizes are illustrative.

Netting and tree stakes	0.9–3.0m long	5–10cm diameter
Fence posts (ordinary)	1.2–2.0m long	min. 6cm top diameter
Strainers/corner posts	up to 2.3m long	min. 12cm top diameter
Gate posts	up to 2.6m long	15–25cm top diameter

Struts and anchoring stakes are typically 2m long and about 10cm diameter. Rails are the appropriate length and at least 5cm diameter.

Other types of stakes will have different dimensions. Supports for tree shelters can be cleft chestnut, rough sawn (3cm) battens, or small stakes (3–4cm diam). Hedging stakes, set every metre through a newly laid hedge at about 80 degrees, need only be 3–6cm diameter.

Cutting and preparing posts and stakes

Stakes usually come from first or second thinnings in plantations or from the tops of trees in later thinnings. Once the felled tree is delimbed, stake lengths are cut according to need. Most stakes, especially from conifers, are used 'in the round'. However, a large diameter stake can be split (cleft) into two, or sometimes more pieces. Traditional oak post and rail fences are cleft as is the special form of chestnut fencing where palings are wired together – and are often seen cordoning off building sites.

Pointing stakes can be done on a sawbench with a circular saw, but do ensure that you are trained and wear protective clothing. If you only have a few stakes to point, then use a hand axe. Make sure the axe is sharp. Hold the stake firmly at about 25 degrees off vertical and rest the end for pointing on an old tree stump, log or discarded plank of wood. Start trimming off the bottom 15-25cm with vertical strikes of the axe to create a taper of about 25 degrees. Once one face is tapered turn the stake a half rotation and taper off the second face. This will leave a wedge shape. Trim the sides of the 'wedge' to fashion the third and fourth faces and so complete the pointing. There's no need for a sharp point, which is easily broken off, a small square end is best.

Preservative treatments

Where non-durable timber is used for stakes or other fencing material that will be in ground contact, preservative treatment is essential. However, simply painting a surface application of a product will not do, despite claims on the container, though diffusion of borate-based preservatives is sometimes possible. For some species the sapwood will retain preservative but the heartwood won't. Some sawmills have on-site preservative treatment plants and may be willing to treat your stakes. It is generally simpler and better for the environment, to use species that have some natural durability like larch, yew, cypress, oak and sweet chestnut.

Note that it is now illegal to use creosote – for me, though, its smell still evokes happy childhood memories of helping my father 'creosote' the garden fence every few years.

A few tips when using stakes

Space does not permit a detailed description of erecting a fence, but here are a few tips from the best of all classrooms, experience in my own wood:

1. Always open up a hole before driving in the stake. You can either use a post hole digger to create larger hole, or a crowbar to create a smaller one. This guides the stake down at the right place, helps keep it vertical, and you discover what the soil is like and whether there are stones. Also, the actual driving in of the stake will be easier.
2. Use a proper sledge hammer, not the back of a 7lb axe. Better still use a stake-driver i.e. a 1m long iron tube sealed at one end and with handles, that slips over the top of the stake. Using an up and down pumping action, the tube's weight and your own force drives in the stake.

3. If the top of the stake becomes badly scuffed or split, cut off the top 1cm or so with a bow saw after the stake has been fully driven home. Alternatively chamfer (trim back, creating a bevelled edge) the edge around the top and splitting will be unlikely.
4. One-quarter to one-third of a stake's length should be driven below ground.

Happy staking.

Using greenwood

Today there is great interest in using freshly cut wood in the round – not sawn – and using unseasoned timbers for rustic purposes or even for buildings. Ben Law's famous house built from woodland produce using traditional skills has sparked widespread interest. There are really two key points to note.

Firstly, one works with nature's natural shapes. You look for a bend or a fork or a certain length for a particular purpose to minimise shaping and cutting the wood itself. Added to this, splitting with the run of the grain is preferred to sawing to help optimise strength as the fibres of the wood are not cut.

The second point is that the wood or timber dries *in situ*. As it dries (seasons) some twisting, shrinking, cracking, warping and other movement is inevitable; changes which can go on for a long time. Some years ago I seasoned a 6ft plank of 3 x 7in oak cut from my own wood to use as a mantelpiece. I was, of course, only using the heartwood. It tested my patience, drying very slowly at about 1in of thickness per year. Only after three years was it sufficiently well seasoned for movement to stop and the mantelpiece put in place for life in the warm and dry!

It's good to work with nature, but recognise you are using a very different product from the dried, planed and shrink wrapped offerings of your local DIY store.

8

Firewood and Wood to Burn

We've added this chapter because firewood has become so much more important in the last 10 years and woodburning stoves commonplace. Also in some circumstances you can obtain grants to support management for fuelwood production or wood burning installations in your home.[1]

If you own a woodland, one of the easiest ways to take advantage of it as a resource is to burn it! Logs are the oldest form of energy, and one of the most renewable. Using them can be as simple as picking up some fallen branches on your way home, or as complicated as detailed yield models, processing and predictions of use. I, that is Will, have aimed to provide straightforward advice with some references to more technical information if you feel that way inclined. I do appreciate that not everyone wants to be as geeky about this as I am!

The art of managing a woodland for fuel is really all about matching up your expected demand against your potential supply. While it is unlikely that you'll need more fuel than your wood can provide (even a single mature tree is likely to keep you going for a good while) it is useful to establish how much you need to fell and split in order to meet your future supply. Wood does take a long time to dry and ideally you could do with planning at least two years ahead.

Faggots once fuelled our hearths, a return
to burning firewood helps the planet

---

1 www.hetas.co.uk

Getting set up

Before we go any further, I should point out that I'm going to assume that you're using an enclosed stove, and not an open fire (or a more sophisticated biomass system). This may seem a bit unrealistic, or even a little unfair, however; there are some very good reasons why. Open fires are shockingly inefficient: they take your log, throw about 80 per cent of it into the local atmosphere, and radiate the remaining 20 per cent into the room. When they aren't doing that, there is a great big hole sitting there that draws hot air out of the room! The upshot of this is that you are probably using significantly more fuel to heat your house, whether that is logs, gas or anything else.

I'm not going to go into intricate detail about how to choose and install a stove, but there are a few things you do need to make sure are in place before you start putting flame to tinder.

Was it installed correctly?

Log stoves are not complicated bits of equipment. They don't have many moving parts, don't require a degree in engineering to understand, and it's not completely outrageous to suggest that you could install your own. The building regulations recognise this and don't specify that you have to have special training or a licence to install; they merely state that the installer has to be a 'competent person'. Having said all of this, it is worth remembering that simple isn't the same as easy and that if you get it wrong you run the risk of a) ruining your stove and b) causing a fairly serious health hazard. The easiest way of finding someone that has the proper training is via the HETAS website[1] and looking at their list of qualified installers. Even if you don't want them to do all of the work, it's probably worth getting someone to check you're doing it right.

Is your chimney lined?

Most modern stoves will be installed with a flue liner. Liners provide a measured diameter to your flue that will help your stove to operate as designed, they also reduce the risk of carbon monoxide (CO) leaks into other rooms (usually bedrooms) that the flue passes through. Finally, a flue liner provides a smooth, insulated surface that will reduce the amount of soot and tar deposited – reducing the risk of a chimney fire.

Do you have a CO alarm?

There really isn't an excuse for not having one of these if you're burning wood (or anything else for that matter.) If incomplete combustion occurs in an appliance, which it will do occasionally even with the best intentions, you will be generating carbon monoxide. This is an odourless, colourless gas that is virtually impossible to detect without the right equipment. It's also toxic; in high concentrations it kills people. I don't mean to scare anyone, and I certainly don't want you to stop using your stove, but a risk exists, and the sensible thing is to minimise that risk. As I said in *The Log Book* (see later for details), it is much easier to install one of these than explain why you didn't to the coroner.

Are you in a smoke control area?

Smoke control areas were a response to the classic London pea-souper smog of the past. In a period where pretty much everyone ran on coal from domestic stoves to heavy industry, poor air quality was a significant health hazard. It isn't a coincidence that the most expensive housing in most UK cities with an industrial heritage is on the western (or windward) edge. Most urban areas in the UK are still designated smokeless zones, where it is illegal to emit (rather quaintly) 'dark smoke'.

There are two ways you can burn a fuel legally in a smoke control area. Either burn an approved 'smokeless fuel' (like gas, coalite etc.), or burn the fuel only in an 'approved appliance': essentially an appliance that appears on the DEFRA list. No form of wood is approved as a smokeless fuel, so you need to make sure that your stove is listed. You can check whether your appliance (or the one you want to install) is listed on the DEFRA website.[2]

It's worth noting at this point that it is also an offence to *supply* non-smokeless fuel in a smokeless area *if you know* that it won't be burned in an approved appliance. I have never heard of anyone being prosecuted under this bit of legislation, and often you have no realistic way of knowing how or where the fuel will be burned, and as far as I can see (though I'm not a legal expert) there does not appear to be a legal obligation on you to find out. However, I suspect (though I can't prove it) that we're due for

2 http://smokecontrol.defra.gov.uk/

a bit of action in this area, as log stoves rise in popularity. This is something to be aware of if you're planning on selling your surplus timber to friends to fuel their fires.

How much fuel do I need?

Unfortunately this is not a straightforward thing to estimate. Unless you've been operating a stove for a number of years, you're unlikely to have any real data to go on. The energy embodied in logs is variable depending on the moisture content (also less so by species – more later) and the amount you need will depend on the weather outside.

It is possible to do some ballpark estimates of the amount of fuel you might need, and then measure how much you're actually using from year to year. I've done a basic calculation based on using a stove for 150 days per year with the stove actually lit for an average of about three hours a day. So, as a rule of thumb:

Burning green wood (around 50% water)	0.28 tonnes / kW of stove output / year
Firewood seasoned for 1 year (35 to 30% water)	0.20 to 0.18 tonnes / kW of stove output / year
Firewood seasoned for 2 years (25 to 20% water)	0.17 to 0.16 tonnes / kW of stove output / year

To apply the rule, simply multiply the figures above by the rated output of your stove, so a 5kW stove will need about 1.4 tonnes of green wood a year, or 0.8 tonnes of really well seasoned wood. You can also play around with these figures a bit, so for example if someone is going to be in the house in the daytime and you estimate that you'll need the stove for nine hours a day, simply multiply the amount of fuel needed by three.

Please do be aware that this is a very rough method, to see the full calculation and understand it fully, see *The Log Book: Getting the Best From Your Woodburning Stove.*[3]

---

3 *The Log Book: Getting the Best From Your Woodburning Stove*, Rolls. W, Permanent Publications, 2013.

What does that look like?

Unfortunately (again) trees are not a simple size or shape to measure, particularly when they are growing. There are a number of methods for measuring, but they grow increasingly more complex as accuracy increases. Volume of course is not the same as weight or moisture content and these variables conspire to make the process quite difficult.

I have provided a few more assumptions here that I use when I'm trying to work out if a project is plausible. They are *very rough* rules of thumb and are nothing like accurate enough to pass muster with any timber buyers or contractors. What they do however is give you a rough and ready idea of the order of magnitude that you're working to. When working with a small log stove, the quantities involved are usually quite small and your decision making process is likely to be whether to fell one tree per year or two. You will probably find that you get a good feel of the sorts of quantities you're after relatively quickly, but if you want to be scientific about the process, there is a detailed methodology in *Forest Mensuration: A Handbook for Practitioners.*[4]

Volume to weight

One tonne of freshly cut timber is about equivalent to a cubic metre.

Annual increment

Increment is a measurement of the amount of new solid wood produced by a tree in a year. This is usually averaged over an area of woodland and the expected lifetime of the tree. A conservative estimate is that coniferous woodland is capable of achieving at least 8 cubic metres per hectare per year (m³/ha/yr) on most sites and often much more, while broadleaved species are capable of producing around 4-6. Bear in mind that this will vary considerably depending on site conditions, management history, the pressure on trees from pests, diseases and pollution etc. You'll also need to remember to take into account any open spaces within the woodland.

This is a guide to the average annual growth of an *established* woodland and doesn't take into account the length of time since

---

4 *Forest Mensuration: A Handbook for Practitioners*, Matthews R. & Mackie, E., Forestry Commission Publications, HSO, 2006.

the last felling. If you haven't felled any timber for some time, there will be a larger stock available in the woodland for the initial cut. Again I'm including this to give you an idea of the sorts of amounts it's reasonable to expect, rather than providing a precise estimate.

Stacked volume

Logs are awkward shapes and sizes and don't pack conveniently. You will always have air gaps between logs and this will vary depending on the size, neatness of stacking, and variation in shape. A useful conversion is that $1m^3$ of solid timber is roughly equivalent to $1.5m^3$ of neatly stacked logs, or $2.5m^3$ of loosely heaped logs. Knowing this helps you visualise the size of the woodpile you'll need to see you through the winter.

Turning a tree into fuel

Choosing species

There is a lot of information flying around that purports to tell you which species are best for burning. This ranges from very authoritative sounding guidance from old hands in the stove industry to a couple of poems that regularly do the rounds. The fact of the matter is that there is virtually no variation between different species in calorific value when measured by weight. There is greater variation when species are measured by volume as (predictably) the denser woods have a higher calorific value per m^3. This variation in calorific value, however, is still substantially less than the change caused by variation in moisture content. Incidentally, this is why ash has such a good reputation as firewood: it has a very low green moisture content (around 33%) which means that it has a higher calorific value when green, i.e. straight from a felled tree unseasoned, than other species. As soon as you dry them all down to the same moisture content, this advantage disappears.

My main advice is: *burn what you have*. As long as you dry wood thoroughly and burn it in a well maintained stove, pretty much any species will burn cleanly and hot. You may find with experience that you prefer one species over another; this might depend on how easy a species is to cut and split, or the smell, or the texture. Personally, I really like birch as a firewood, and some of the softwoods, I also hate the smell of holly and elder – but it really is up to you – it's your wood!

A quick note on poisonous species

1. During my career, I've been asked many times whether it is safe to burn many different species. This has ranged from some that are poisonous, to others that are really quite harmless. The answer, as always, is it depends. Complete combustion should make any organic poisons inert, breaking them down (usually into carbon dioxide and water and not much else). However, complete combustion is virtually impossible to achieve in a log stove and is completely impossible in an open fire. This means that while any vapour or particulates coming off the wood are likely to have decomposed in the heat, it is impossible to state categorically that they have done so.
2. Flue gasses are not something that you want to inhale in high concentrations regardless of the species of wood that is burning at the time. This is particularly true if you are burning coal as well. Wood stoves are enclosed units and you shouldn't be getting flue gasses escaping into the room (see the section on page 97 about carbon monoxide) but if you are using an open fire, then this is going to be a consideration.
3. Green wood is more likely to produce vaporised sap than properly dried wood, as is foliage.
4. I have contacted the poisonous plants information service at Kew, and while they don't have definitive answers, they do have anecdotal evidence that cherry laurel and rhododendron can cause problems, because of vaporised sap while processing (though **not** while burning).

Given these points, my personal opinion is that as long as you're burning seasoned wood, hot and fast in an enclosed stove, you're unlikely to have any problems. I have certainly never heard of anyone suffering from ill health due to the **species** of wood they were burning. Ultimately, you need to make this decision for yourself.

Processing choices

There are some tradeoffs to be made in which order you cut, split and dry.

Change in hardness and ease of splitting
As wood dries some shrinkage does occur which can change how easy the wood is to process. Some woods (willow) are so wet and spongy when green, that if you hit them with an axe it just sinks in and spits water back at you, and so are probably better dried first. With wet sycamore the axe may sometimes bounce off. Other woods such as oak are already quite hard to split and become even harder as they dry. Occasionally you may find that you've got some wood that fractures very easily when dried, this is due to stresses that build in the wood as it grows and then shrinks while drying, but it tends to be on a tree-by-tree basis rather than predictable by species.

Drying time
The rate of drying is directly influenced by the distance water has to travel before it is able to evaporate away from the log's surface. Larger logs will take longer to dry, and so will logs that still have the bark attached. The bark is an effective barrier to moisture (one of the functions it provides in the living tree) and a log with a split face will dry much more quickly than a log of equal size that still has bark on.

It is worth remembering that dried wood will be noticeably lighter than green material, so it's often worth waiting until the wood has dried before you transport it out of the woodland. Another thing to remember is that if you have large amounts of branch or coppice material available, you may be able to get away without splitting at all, but it will take longer to dry (though I think that'd be a fairly small price to pay for removing such a time consuming job).

Theft
Sad to say, the popularity of log stoves recently has lead to an increase in thefts of logs from woodland. It is much easier to transport cut and split logs rather than whole tree lengths, and this could be a problem for you, if you're near some convenient

access, or a public right of way. If you leave cross cut and split logs visible, do bear in mind that even if it's not accessible to steal, there is always the possibility that some dangerous clown might think it'd be fun to set it on fire instead.

Drying

Length of time
While wood can be burned at any time post felling, the sensible thing to do is to dry it thoroughly before use. This maximises the amount of energy you get out of the fuel and ensures that your stove will burn as cleanly as possible. The rate of drying is variable and depends a lot on the size of log, but also on the weather. If we are lucky and get a long dry summer, you will probably manage to get wood dryer than the figures I've given below, though with our usual monsoon, the ambient humidity will be much higher and reduce the rate of drying. I've given some indicative moisture contents and drying durations below:

	Moisture content	Description	Indicative kWh per tonne
Green	Around 50% water	As it grows in the living tree	2,300
Part seasoned	30-35% water	Around one year drying	3,190 to 3,490
Fully seasoned	20-25% water	Around two years drying	3,790 to 4,090

As you can see, there is a substantial rise in the amount of useful energy you can get from a log when it is fully dried.

The drying process
In the last few years, a Norwegian national television station broadcast a seven hour program on the subject of burning wood correctly. It sparked a national debate. Not, as you might think, on whether it was a good idea to broadcast such a long

programme, but on whether it was better to stack logs bark side up or bark side down. I am not going to court controversy by suggesting one is better than the other, you do however, need to make sure that a) your stack is able to breathe and b) it is not going to re-wet from any nearby source. Dry wood acts just like a sponge and will happily reabsorb water when it's allowed to.

To prevent re-wetting, you need to take into account water moving up from the ground and down as rain or snow. Ideally you want to make sure that your stack is on bearers (a sacrificial layer which you're not going to burn) that prevent moisture being attracted to the bottom of the stack. These can be any old bits of wood like fence posts with preservatives in, or old pallets. Essentially you want to ensure that the fuel is not able to wick moisture up from the ground, and that you have a free flow of air under the stack to allow any moisture that does make its way underneath to evaporate.

Preventing re-wetting from above is a fairly straightforward process of getting some old fence panels or tarpaulin and covering the top of the stack. Remember that you want free airflow so don't bother covering the sides tightly as any airtight surfaces will have a tendency to sweat in the summer.

You can test how dry a log is in a number of ways, but the easiest way of seeing whether it is dry enough to burn is to look for radial cracks in the wood and bark that flake off easily. If you want to get a bit more sophisticated about the process, you can use a moisture meter. Don't worry about getting an expensive one, as price is not really a good indicator of accuracy, but you will need to sample the logs correctly to get a good result. You should check several logs from different points in the stack, size, and species (if present) and take an average. When testing a log, you should split it first and check the moisture right in the heart of the log, where the moisture content is much less variable than on the surface where it can be affected by the recent weather.

Stack location
How you place your stacks will depend on the order you intend to process dry and transport your logs. The objective is to gain the maximum benefit of the sun and wind in drying the logs, while minimising the risk or re-wetting, theft etc. and the inconvenience of transporting it more than necessary. You'll

also need to take into account the space required for storage and processing. There are many, many different methods and tactics that people use to ensure a good supply of dried logs at the point of use, and you are probably best to do some experiments yourself to find out what works best for your situation.

Processing

The methods of processing vary depending on the quantities you need to work with, the amount of time and energy you're prepared to invest and the type of wood you need to process.

Processing by hand

At its most simple, this is a bow saw, a log splitting maul and a lot of energy. It will certainly give you a good workout and is probably the cheapest and simplest form of processing. I quite enjoy this (because I get to hit things hard with an axe) but I suspect that its appeal will wane somewhat as I get older. If you're only producing small amounts for a stove you use occasionally, then this may be the most effective option.

Motor-manual (using a chainsaw)

This is essentially the same as processing by hand, only with mechanical advantages. Using a chainsaw and hydraulic log splitter will speed up the processing and reduce the amount of hard work required (a bit). It will certainly enable you to get through greater quantities than you would by hand, and will allow you to deal with more awkward shaped lengths than a fully automatic processor. I may be criticised for stating the obvious, but please don't try and use a chainsaw if you haven't been properly trained, and certainly not when you are on your own. I can't think of a more dangerous device that you can buy over the counter in the UK without a licence, no questions asked. If you doubt my words (and you have a strong stomach) then just try typing 'chainsaw injury' into Google. There are guides about how to carry out forestry operations safely on the HSE website,[5] and you can get information about the proper training from Lantra[6] (the sector skills council).

5 www.hse.gov.uk
6 www.lantra.co.uk

Fully automatic

Log processors are available that produce very high volumes, automatically cross-cutting to a pre-set length and splitting into a container. These machines work best with straight lengths of timber that are a fairly consistent diameter (usually conifer thinning). Any material that is too large (such as large broadleaf stems) or crooked will not feed correctly and need additional attention, which rather defeats the object. You are very unlikely to need one of these to produce logs purely for personal use, but if you have neighbours who produce logs then you may be able to borrow or hire one for short periods.

Selling the surplus

If you find that you have large quantities of wood that is too crooked to sell to a sawmill, you may find that small log sales are a potential market. The price that logs fetch is notoriously variable depending on a range of factors including the weather, local competition, species and delivery distance. Log suppliers often deliver by the 'load' (whatever that is) that muddles the picture further. Before you invest in any machinery, I'd suggest doing some homework on the other suppliers in the area and the going rate for fuel grade logs. This should give you an idea of whether you want to produce yourself, or sell surplus timber to other local contractors.

Using logs can be as complicated as you like, but I hope you enjoy your burning!

9

Enriching Your Wood for Wildlife:
Some Practical Tips
to Increase Biodiversity

One of the joys of trees and woodlands, whether your own or in the countryside at large, is that they add hugely to the richness of wildlife. There are several reasons, both familiar and perhaps surprising. Long-lived trees, and stands of trees in woodlands, add structure and variety of habitat, almost a third dimension compared with a field of wheat or grassland, and they offer countless niches.[1] Also, woodlands are often some of least spoilt elements of our natural heritage with many of them, known as ancient semi-natural woodlands, always having been under trees and so providing a direct link to the ancient 'wildewood' that once covered two-thirds of Britain. And woodlands, compared with farms and gardens, rarely have pesticides, fertilisers, or other chemicals inflicted on them in the course of management: or if they do it is only once or twice in 50 or even 100 years at the time of initial planting. Often rides in woodlands, having never been cultivated or ploughed up, can in the less shaded parts be refuges for wild flowers of meadow and hedge bank. Woodlands are havens.

An excellent guide for what to do in a smaller woodland is Blakesley and Buckley's short book, *Managing Your Woodland for Wildlife*. (See appendix on page 155.)

If there are such benefits, how can one make the most of them? How can tree and woodland management enhance wildlife? We will try and answer these questions by looking at three kinds of sites, though in each case practices for one will usually be applicable to one or both of the others.

Before we start, a brief comment on why complete neglect

---

1   Ecologists use 'niche' to describe the type of home or environment an organism needs in order to thrive.

isn't often best for wildlife. Quite apart from the danger of leaning trees, fallen branches and other hazards, years of neglect cause rides to become filled in, glades overgrown, aliens such as grey squirrels and muntjac to run amuck browsing or gnawing everything in sight, or non-native trees such as the common Turkey oak or the invasive western hemlock to dominate. I exaggerate, but only a little. The point is that frequently the net effect of neglect is less variety in structure, fewer wild flowers on the woodland floor, and overall less rather than more diversity.

Ancient woodlands

Ancient woodlands are more properly called Ancient Semi-Natural Woodlands (ASNW) since all woods in Britain have been disturbed to some extent by man, by operations such as coppicing and pollarding. However, they are often the most valuable woodland type for wildlife. This is because the land has always had woodland cover and thus usually possesses a flora that is both rich and also often of plants only found in such places. 'Always' is actually defined! Land believed to have been under woodland since 1600 (1750 in Scotland) is likely always to have been wooded; that is how 'ancient' is defined. The countryside agencies like Natural England, Natural Resources Wales, and Scottish Natural Heritage, maintain details county by county of all woodland considered to fall into this category of ancient.

If you haven't checked before, do find out if the woodland you own is ancient. You will usually be blessed by abundance and variety of wildlife – some ASNWs have over 200 species of flowering plants alone – and also 'blessed' with a little extra red tape with what you can do with it!

The lovely yellow archangel often indicates an ancient woodland site.

Two of the features of many ASNWs point to ways that all woodland can be enhanced for wildlife.

- Firstly, many such woods have mature, over-mature, dying and fallen trees which provide countless niches. Holes in trees for nesting and for bats are an enormous boon and fallen, rotting wood – dead wood – provides habitats for countless invertebrates. Such conditions are rarely found in the plantations of tidy minded foresters, not to mention one's own garden. Deliberately encouraging such tree and woodland conditions, by extending rotations and not rushing to clear up fallen timber, are useful ways to promote biodiversity.

- The second feature is that many ancient woodlands will have been managed in traditional ways. Today's preponderance of growing trees all together to large size, 'high forest' as it is called, was the exception in the middle ages. Most woodlands, as we mentioned in an early chapter, were kept quite low by coppicing to provide the small sized wood products then in demand. Resumption of these traditional practices, mainly coppicing and pollarding, keeps alive the wildlife so long associated with them. The influx of light and warmth to the forest floor stimulates dormant seed to germinate and results in the magnificent show so often encountered in the years following a coppicing. It is one of the glories of England! So exclaimed that irascible, itinerant farmer and politician of the early 19th century, William Cobbett, in his *Rural Rides*, 'What in vegetable creation is so delightful as the bed of a coppice bespangled with primroses and bluebells?' The regular and cyclic cutting of coppice allows such flowers and the associated wildlife to flourish. The clearing of the ground, the letting in of light, and the replenishing of the seed bank in the soil every 15 or 20 years, creates this glory of God's creation, artificial though the management is.

If coppicing is resumed plan to do it in a succession of areas at intervals of a few years. Butterflies, such as some fritillaries, thrive only in recently cut areas so to maintain a population one needs sunny glades, freshly cut coppice and for tracks and rides to be the corridors that connect them.

There is of course, much more to ancient woodland, but highlighting some of the reasons why they are rich and good for wildlife helps us to know how we can make use of such practices more generally.

Recent woodlands

Recent woodlands are ones known to have been established on bare land. This is the case for much of my own wood that was first planted in the 1880s on former farmland. However, before turning to these, some readers may be wanting to remonstrate with me for not commenting on a third key feature of ancient woodlands, that is they are almost exclusively of native tree species. This is undoubtedly correct, but as a reason for their being rich in wildlife it is less significant than is often suggested. The late Sir Richard Southwood's seminal work in the 1960s showed a significant correlation between variety of insects and tree species, in particular how long the species was believed to have been part of Britain's flora. Oaks, an early arrival, possess hundreds of associated insects while recent introductions, such as many of our commercial conifers, only 30 or 40. However, this genuine correlation has many anomalies. For example, native beech and ash are inferior in insect species diversity to some recent exotic introductions. Indeed, assessments show that the southern beeches (*Nothofagus* spp.) from Chile and Argentina, which have only been planted in the British Isles for about 100 years, support far more diverse populations than common beech, and are only exceeded by oak itself. And talking of common beech, it is thought with good reason only to be native in Britain south of the M4 corridor, though not as far west as Cornwall, possibly in Essex, and also in the county of Gwent in SE Wales. Some question whether it is native at all, but rather a very early introduction in Bronze age times.

The point is that far more important than tree species appears to be structure in woodland. It is uniformity that is unattractive to wildlife. Each of the following adds diversity:

Glades

Ensure that woodland has some glades open enough to encourage sunlight on to the ground. Around these glades shrubs will grow and this will increase habitat known beneficially as 'edge effect'.

Woodland glade showing sunny open area, shrubs, and tall trees which is ideal for wildlife and pretty good for a picnic site too.

Rides

As with glades, make sure some rides are open and allow plenty of light and warmth. They will be sunniest if running East–West. Don't cut both sides of a ride every year but alternate the side for cutting to allow ride-side plants to flower and set seed. If there are no glades, add them to rides by opening up a bay every so often that can double for wildlife and somewhere to store cut timber or as a campsite. Make such 'scallops' about one tree height distance into the adjoining stand.

Ponds

If your wood is without open water why not consider constructing a pond? While this is not the place for great detail, the key elements are:

- To excavate on a flat site to a depth of 60 to 90cm;
- To have full overhead exposure above the pond itself for sunshine and so that overflying birds can see it;
- To have a least one bank that gently slopes to provide some shallow water;
- To avoid a very regular shape but include an 'inlet' or even a tiny island;
- To line with a proprietary butyl liner or, if you can, to puddle with local clay;
- To check for permissions if you plan to abstract water to fill or maintain it.

In my own case I let the pond fill up from rain. This worked well, but in the long run I was thwarted by other intruders (man and animal) who pierced the lining many times, but that's another story!

Thinning
Neglect of thinning is the scourge of many woodlands. Thinning out poorer trees will help the remainder to grow better and will open the canopy allowing sunflecks onto the woodland floor. It transforms a dark stand of trees into a more open and usually beneficial environment for wildlife. You don't need to thin out every unwanted tree. Any that are not interfering with good trees you can retain. In the beech stands in my own wood I have left many small, suppressed trees to create patches of two storey forest just to add structure.

Sow wild flowers
Recent woodland is often impoverished and the addition of common wild flowers such as scabious, bellflower and red campion, can be achieved by sowing seeds. Research in the 1980s by Dr Joanna Francis and others has shown how successful this can be. Visit the new woodlands around the new town of Milton Keynes and you'll be surprised by the displays of flowers on the unpromising clay soils. Buy seed, plants or bulbs from reputable suppliers able to provide material of known British origin.

New sites

As a reminder of what we said before about planting a new wood, lay it out so that wet areas, stream sides, rocky sites and other interesting wildlife-rich features, including bits of woodland, are left untouched. Retain hedges, ditches and banks and, of course all archaeological features. If there are mature trees, retain these as well. In addition make provision for wide rides and glades right from the start. Plant native tree species if you want to, but also add some shrubs such as hazel, hawthorn and even spindle tree with its lovely November display of tiny pink and red fruits.

Trees

First, health and safety

For much wildlife, the ideal is for trees to grow as large as possible and live for as long as possible, but this can lead to unacceptable conditions for safety! You can't have a large, old tree decaying and falling apart right next to your entrance or overhanging a road where it is a hazard. And the expense of employing tree surgeons to reduce or thin dangerous crowns to prolong a tree's useful life is usually out of the question. Once holes develop in trees they often indicate decay within and increased hazard without. However if the tree represents no threat, leave it for the abundance of nesting sites it provides, the preserve of beetles, and the substrate for fungi and micro-organisms of all kinds. If it must be cut, you can always put up nest, bird or bat boxes, leave the trunk as a rotting hulk and the branch wood as deadwood habitat on the ground.

Nesting box suitable for tits. Remember to clean out the box in early winter.

Dead trees and snags

Once a tree dies it doesn't mean the end of its usefulness for wildlife. Indeed, while it stands – what Americans call 'snags' – raptors (birds of prey) will use it as a perch, owls will nest in cavities, woodpeckers will seek out insects, and those lower down the food chain will benefit in consequence. When safe to do so, and when not wanted for other purposes, leave dead trees standing.

The green woodpecker and the tawny owl are just two of the many birds that value dead and dying trees full of holes and cavities.

Ivy

There is no need to cut ivy away from a tree. This climber provides splendid cover in winter and whenever possible should be retained. It seldom harms the trees it climbs up, though if the tree is already weakened the extra 'sail area' offered by an ivy-clad trunk may expose it more to storm damage.

Ivy has another boon. It is pretty well the last wild plant to flower in our countryside, typically November and early December, and provides a late season nectar source for bees. Such bee 'pasturage' helps their energy reserves for over-wintering.

Resuming pollarding

Sometimes old and long neglected pollards can be brought back into cycle, but the key appears to be to leave one living limb attached while the recovery phase lasts. Cut back branches to just outside of where they were last removed and hope that adventitious buds (new buds which often develop in the callus tissue at a cut) will do the rest. Success is not guaranteed. If it does not work, you still have a large trunk that becomes a 'snag' and will benefit wildlife in countless other ways.

Conclusion

To sum up for woodlands generally, wildlife is helped by providing diversity of habitats by:

- Conserving existing natural features;
- Allowing trees of all ages and conditions;
- Avoiding uniformity and encouraging varied structure;
- Resuming practices such as coppicing, pollarding and thinning;
- Providing areas of light and dark – glades, open rides;
- Leaving some dead trees standing and creating deadwood piles;
- Thinking about deliberately adding wild flowers in to recent woodland;
- Excavating a pond.

Do all the above and you will be a blessing to our wildlife and to all of us who enjoy the countryside. And since badgers love an undisturbed corner of a wood a few yards in from a field for their sets and love mushrooms and toadstools for food, they may even turn up as a 'thank you' for all your hard work and your blisters!

Food for thought: a badger eyeing
a cluster of pixy-caps.

10

Keeping Your Wood Safe

Owning or looking after a woodland is usually a long term commitment. As the owner of a woodland you will want to ensure that it thrives for future generations to enjoy. Part of your role will be to safeguard and defend your wood against a range of different factors that can harm trees and prevent new growth. These factors have different levels of severity and likelihood, and there are different things you can do to prevent or mitigate the risk.

I (Will) don't expect you to read this chapter from beginning to end, as there is quite a lot of information to take in all at once. For each factor I've given a brief overview, and where possible links to further information, and where to get professional advice. I'm afraid that this chapter might seem a bit doom-laden and gloomy. It can get rather depressing when you can see all of the different ways your woodland can fall apart or die. It is quite important to retain a sense of perspective when we're looking at all of these problems. They rarely all crop up at once, and many carry such a low risk that you can happily ignore them. The fact that we have so many woodlands, and that the woodland area in the UK is actively growing is cause for optimism. Humans may be playing in the woods, but the trees are serious about their battle for survival, and it's one that they have been mostly winning for thousands of years.

You can divide the risks up in many different ways, but the simplest is to look at them in terms of the root cause: physical factors, such as climate or vandalism; chemical factors such as soil contamination; and biological factors (which are usually the most pervasive) caused by pests and pathogens.

I have deliberately stressed issues and assumed you have some intention of managing your wood for a commercial return. For many readers this will be unimportant, though a number of the biological factors will have a severe impact on your woodland regardless.

Interactions

Trees are typically quite resilient to a range of factors, and while many of the things that can cause damage are impossible to prevent altogether, trees that are under stress from one or more elements, will tend also to be more vulnerable to other threats. The thing to bear in mind about all of these problems is that many of them are limited in scope when viewed in isolation. The thing that causes the most damage to a tree, may not be the thing that actually kills it: often the final coup de grace may be delivered by a particularly bad storm or heavy snowfall, but the underlying problem has been caused over a number of years by other stressing factors which make the tree more susceptible to damage.

Responses

There are a range of actions you can take in response to the threats I've outlined here. Some of them may only be a threat to a particular species, or type of tree (seedlings for example), and some of them may warrant closer attention or long-term actions.

Avoid
Avoiding the problem is often the best way of dealing with it. You may find that you can sidestep many of the risks to woodland by making careful silvicultural choices – diversifying your species mixture for example, or by making sure that you've done your homework on local threats properly beforehand. Another way of looking at it would be refusing to import the problem in the first place. Many diseases are brought into woodland on seedlings from infected plant nurseries. A bit of extra care to begin with can save a lot of headaches later!

Protect
Many risks to woodland are temporary, while trees grow and adapt. Some are caused by occasional nuisances that can be dealt with on an ad hoc basis. Examples of this kind of response include fencing areas at risk of damage from browsing mammals until the trees are big enough, working with local police to prevent

vandalism or poaching, or working with contractors to make sure that damage during site operations is minimised.

Ignore
Some of the factors I've outlined below are very rare. For example, lightning strikes are few and far between (and there's not a lot you can do to stop them). These factors are best ignored, because any work you could be undertaking to protect against them will almost certainly be more expensive than the event itself.

Occasionally, you'll come up against a problem where there is no clear answer. Finances don't permit making big changes to management, or specific factors such as road spray drift just tend to occur in your woodland at a particular time of year. The world is full of foresters who'd just like to do a bit of extra pruning, weed control or fencing, but the budget doesn't stretch that far. At this point, it's better to accept the problem and rather than spending time worrying about what might happen, try thinking of ways you could make your wood more resistant in the future.

PHYSICAL FACTORS

It might appear at first glance that there aren't many things you can do about a lot of physical factors. While it certainly is the case that many of them can't be prevented as such, there are things you can do to mitigate the risk, or allow your wood to adapt gracefully to a gradual change in circumstances.

Climate

Leaving aside the problem of climate change for the moment, there are a number of weather-based threats to woodland. These tend to come primarily in the form of extreme weather, or an impaired ability to cope with extreme events.

Wind
Paradoxically, wind rarely hurts healthy trees that are growing individually, but it can cause havoc to woodland in the right circumstances. It is also a potential threat just about anywhere in the UK and isn't confined to the uplands. Trees grown in open

conditions or on woodland edges tend to become *windfirm* – that is, more resistant to high winds. This resistance is developed because they have been exposed to higher wind speed while growing, stimulating a more developed root system than trees of the same age and species grown in the shelter of the woodland itself. This lack of resistance is not a problem as such, until windfirm trees begin to die, or are removed for other reasons, leaving the main body of the wood vulnerable to damage as the remaining trees are unable to deal with the sudden stress imposed on them. This effect can be especially marked in upland conifer plantations, where it is virtually impossible to create a break in the windfirm edge without substantial damage to the remaining woodland.

A good example of this is in Julian's wood where he thinned a stand of old sycamore and ash coppice and found that the St Jude's storm (23rd December 2013) blew down a further 20 tall trees that were newly exposed.

Some woodlands are particularly at risk of damage from the wind:

- Broadleaved woodlands are particularly at risk to storms while they are in leaf, as the canopy represents a huge sail area with which to catch the wind.
- Unthinned areas of woodland will have an even lower resistance to wind as trees will have had virtually no exposure to wind while growing (and may be propping each other up).
- Some locations will have a greater risk of wind damage than others due to an increased frequency and ferocity of storms. This tends to be upland areas, and areas in the north and west of the country.

Dealing with wind is essentially a case of anticipating damage before it occurs. Trees which are already under stress from biological factors or root damage are likely to be the ones at greatest risk, and you may decide to fell these trees in a controlled way before the wind does it for you. You need to be particularly careful of trees with large amounts of dead wood in the crown, as large branches can drop suddenly. While not a risk to the trees in the wood as such, this is a significant risk to anyone in the wood at the time. The opposite approach to this would be to delay felling windfirm trees in a risky area as long as possible to minimise risk to the surrounding trees.

Rainfall

Woodlands are typically quite resilient to heavy rainfall, and can provide substantial benefits to those living downstream by reducing the risk of erosion and flooding. Wet woodland is a particular habitat type that is underrepresented in British woodlands and can be a haven for local wildlife. You can however, run into problems with drainage. These can range from problems with access to the woodland and bogging down your car (or a contractor's machinery) to more significant problems with trees dying from waterlogged roots.

If you have drains in your wood, it would be a good idea to find out where they run, and whether there are any blockages. You may choose to allow drains to deteriorate, to encourage different species to grow, but it would be a good idea to establish how the trees on site would react to an increase in ground water. Waterlogging commonly kills off root systems unless a tree is particularly adapted, this will tend to make a tree far more vulnerable to other stress factors and will usually cause dieback. Some species, such as willow and alder thrive in waterlogged conditions; others such as hornbeam, ash and some conifers will tend to die back quite rapidly.

The opposite problem to excess rainfall is too little. Most UK native species are adapted to handle our usual climate that rarely includes prolonged periods with no rain. There really isn't much you can do to combat drought except to make sure that your trees are healthy and not under stress from other factors. Some of the symptoms of drought are dieback and premature leaf loss. These don't necessarily mean that the tree is a lost cause as these responses tend to minimise the transfer of moisture of the tree by reducing the rate of transpiration.

Frost / Snow

If you have a healthy woodland composed of native species, then frost and snow are likely to be only a minor nuisance at worst. They may stop you being able to plant new trees when you'd like, and you may find that a late frost has an effect on fruit bearing trees. However if you have trees that are particularly old and have rotten or dead limbs then snow may cause them to break, and if you have species that are nearing the limit of their natural range, then they may suffer. As with most other physical factors, there isn't much you can do about it, except over an extended period of time set about changing the composition of the woodland to be more resilient.

Fire

Lightning strikes notwithstanding, the vast majority of fires in woodlands in the UK are started by people. The likelihood of it causing significant damage is fairly low, but some types of woodland are more vulnerable than others. In particular, areas of scrub, heather and heath tend to dry out in the summer, and young un-thinned conifer stands tend to have thick, dry areas of leaf litter that will catch readily. Early spring is also a hazardous time for very young woodlands when grass and herbage among the trees are dead and dry from the previous years and before fresh, new and much less flammable growth has come through. If you think your wood may have a high risk, the Forestry Commission has a free toolkit for assessing and managing the risk available from their website.[1] A research technical note on fire suppression is also available.[2]

People

I have a confession. When I was a student, I accidentally marked the wrong trees for felling. In that situation, I was probably the single biggest threat to individual trees in that plantation: the living embodiment of a little knowledge being a dangerous thing. As usual with any natural system, people are the biggest cause of damage, whether that be deliberate arson, vandalism or accidental damage by contractors (and misguided student foresters). Julian has covered some of the responses to anti-social behaviour on pages 31-34 but it is also worth noting that it is entirely possible that you can be your own worst enemy. Forestry is a long-term business and there really are very few jobs that

1   www.forestry.gov.uk/fr/INFD-7WKJDJ
2   www.forestry.gov.uk/pdf/FCTN3.pdf

cannot wait until you've had a proper look at what needs doing and checked it thoroughly. The old saying of 'measure twice cut once' applies just as much in forestry as it does in carpentry!

Don't use trees as fence posts!

It is fairly common practice in the agricultural community to use trees as living fence posts. While this is not often a substantial threat to the tree itself, it does make the tree very dangerous to fell. Chainsaws, fencing wire and staples do not make good friends! Even if you do manage to fell one of these trees, the contamination has probably made the tree unsalable – as sawmills are usually reluctant to risk expensive band saws. You may deem the price of a few trees worthwhile, if the boundary is particularly hard to fence (and you need to keep the sheep out) but I wouldn't recommend that you do this unless you have to.

Soil erosion / compaction

In a woodland, soil is the most valuable resource. It is the medium that sustains the trees and in the case of semi-natural woodland retains a signature of the collection of plants unique to the site. Removal of the soil from the site, usually in drains and streams, reduces site fertility and can lead to damage of aquatic habitats downstream. The critical thing to remember about erosion is that bare earth is vulnerable, whether it is on sites that have recently been felled or on tracks and woodland rides.

The opposite of erosion is compaction. This is where the soil loses all structure and is beaten down into a hard crust (when dry) or a sloppy mess (when wet). The effects of compaction are broadly the same as for waterlogged sites. Tree roots are unable to respire and typically die, or grow less healthily in the compacted areas. You are likely to see evidence of compaction on forest roads, and to a lesser extent on footpaths.

Climate Change

The elephant in the room is, of course, climate change. The climate that woodlands are so well adapted to is changing. While there is still a lot of uncertainty about how the climate will change and the sorts of issues that will crop up, there is a substantial amount of research about how we can best support woodlands through the transition. The current guidance largely focuses on building woodland resilience by diversifying the number of species and the provenance from which they come. Forest Research has a large body of guidance and further information on the subject.[3]

CHEMICAL FACTORS

Chemical pollutants can arrive in your wood from a variety of different sources, with different levels of controllability and severity.

Airborne pollution

This is another example of a problem that realistically can't be managed as such. This can include acid deposition from sulphur and nitrogen emissions upwind, ozone and particulates from road traffic and spray drift from gritted roads and agricultural operations. Some plant species are more sensitive to airborne pollution than others, in particular lichens are affected very seriously, so if you have a thriving lichen population, the chances are the air quality on site is good.

If your woodland is suffering from airborne pollution, it is worth investing in a copy of *Forestry Commission Handbook 5 (Urban Forestry Practice)*. The handbook contains a very useful table showing specific species that are most resistant to 'smoke and fumes', covering the majority of airborne pollution, and 'seaside' sites, giving an indication of the species most resistant to saline deposition from roads.

Runoff

Groundwater contamination is another insidious problem. It can be caused by industrial processes further upstream, road drainage, landfill sites, and high levels of nitrates from neighbouring agriculture. The clearest indication that you may have a problem

---

3 www.forestry.gov.uk/website/forestresearch.nsf/ByUnique/INFD-7K9DFZ

is likely to be the health of the ground flora near watercourses. The levels of pollution required to kill trees outright are usually very high, but you may find that the additional stress causes increases in fatalities from other sources.

Spills

Whenever you are using machinery and chemicals on site, you run the risk of spilling something noxious. This can range from hydraulic oil and fuel from machinery, to pesticides and herbicides. Whenever you have work going on in the woodland, you should make sure that either you or the contractor has a spill kit handy in case anything goes wrong. Also, some products are manufactured to be biodegradable and, although more expensive, you can, for example, buy such chain oil for a chainsaw and thus be much less polluting.

BIOLOGICAL FACTORS

This is where the worry really begins. Well, perhaps. There are a wide range of pests and diseases that can affect your trees. These vary in severity and likelihood and commonly have different effects on different species. I've covered the main causes of damage here, but the situation is changing all the time, and I've included some sources of further specific information for you to check if you have concerns.

Biosecurity

A term you may come across when discussing infectious diseases is 'biosecurity'. This term (short for biological security) refers to measures taken to protect, or keep secure, one group of biological organisms – in this case trees, woods and forests – from other, harmful biological organisms, such as bacteria, viruses and fungi, certain insects, and invasive plants and animals. There is a guide to biosecurity best practice available from the Forestry Commission.[4]

---

4   www.forestry.gov.uk/pdf/FC_Biosecurity_Guidance.pdf/

Mammals

Mammals are probably the easiest pest to identify, if not by being able to spot them directly, then by droppings, fur/hair and the kind of damage they cause.

There are plenty of different species in the UK that can cause damage to trees. This can be anywhere on a range from the edible dormouse to feral boar, deer and domestic livestock. Damage to trees from mammals usually falls into two categories: browsing (eating leaves, and buds) and bark stripping. Earlier Julian covered the bark stripping problems by the wretched grey squirrel (pages 57-58). As you might expect, there are a number of different methods of controlling these species. These depend on the species, your budget, your personal ethics and scale of the damage caused, but will usually involve either presenting a barrier to the species concerned, such as a fence (pages 56-57), or performing some kind of pest control such as shooting.

The Forestry Commission have a very useful set of resources available on their website. These include a guide to determining the cause of damage to trees[5] and a guide to the options for limiting the damage[6] as well as a number of extended reading lists if you're experiencing particular problems.

Insects

The number of insect species that benefit from trees is far, far too long to list here. Oak, willow and birch alone have well over a thousand insect species associated with them. There is however, a shorter list of species that cause substantial damage to trees. The most common species are listed here:

---

5   www.forestry.gov.uk/fr/INFD-6K4KAF
6   www.forestry.gov.uk/fr/INFD-6LKBAD

Species	Notes
Great spruce bark beetle (*Dendroctonus micans*)	A now well-established pest affecting spruce trees, beetles having been accidentally introduced from continental Europe. They congregate in large numbers and eat the under-bark tissue. Currently the subject of moderately successful biological control. **Further information:** www.forestry.gov.uk/fr/INFD-6XPC8D
Horse chestnut leaf miner (*Cameraria ohridella*)	Causes leaf browning in July and defoliation of horse chestnut trees. While not a primary cause of tree death, may well cause the tree to succumb to other diseases. Present in most of England. **Further information:** www.forestry.gov.uk/fr/INFD-68JJRC
Oak leaf roller moth (*Tortrix viridana*)	When numbers are high, causes damaging defoliation of oak trees in early June. Severe defoliation, while not a primary cause of tree death, may well cause the tree to succumb to other diseases. **Further information:** www.forestry.gov.uk/fr/infd-7b3d8v
Oak pinhole borer (*Platypus cylindricus*)	Tunnels into wood where it infects the tree with fungal spores, which damages the tree's health as well as reducing timber value. While it does occur on oak trees, this species will also attack other hardwoods notably sweet chestnut and beech, but also ash and walnut. **Further information:** www.forestry.gov.uk/fr/INFD-6ZPEJW
Oak processionary moth (*Thaumetopoea processionea*)	The caterpillars can cause serious defoliation of oak trees. Oak is their principal host, but they have also been associated with hornbeam, hazel, beech, sweet chestnut and birch, on the continent. This usually only occurs when there is heavy infestation of nearby oak trees.

Species	Notes
Oak processionary moth (*Thaumetopoea processionea*)	The caterpillars have hairs that carry a toxin that can be blown in the wind. This causes serious irritation to the skin, eyes and lungs in both humans and animals. They are a significant human health problem when populations reach outbreak proportions. So far confined to Southwest London. **Do not attempt to handle the caterpillars yourself, or disturb their nests. If you think you have the pest, contact Forest Research, or your local tree officer.** **Further information:** www.forestry.gov.uk/fr/ INFD-6URJCF
Pine weevil (*Hylobius abietis*)	The pine weevil is a common cause of death in young conifers. Typically beetles feed off the bark of the stem resulting in ring-barking and death. Insecticides are usually used to prevent infestation. Forest Research has much information on their website about effective management and minimising the use of chemicals. **Further information:** www.forestry.gov.uk/fr/ INFD-62WKG9

While I've covered the most common (or dangerous) insect pests here, there are a number of other species that have been spotted in the UK in limited areas. Further information on these potential pests is available from plant health based at the Forestry Commission. They also issue information in species that are not yet present in the UK, but may pose a threat in the future.

Fungal, bacterial and viral diseases

Diseases caused by fungus, bacteria and viruses have been very prominent in the news lately. We have seen an increasing rate of infection of British trees by introduced pathogens, which is very worrying.

Disease	Notes
Acute oak decline	A relatively new condition of oak trees in Britain. It affects older, even mature trees causing black weeping patches on stems with lesions and necrotic tissue underlying the bleed points. **Further Information:** www.forestry.gov.uk/fr/INFD-7UL9NQ
Bleeding canker of horse chestnut	Bleeding on the trunk and branches of horse chestnut caused by the *Phytophthora* fungus. **Further Information:** www.forestry.gov.uk/fr/INFD-6KYBGV
Chalara dieback of ash	A serious disease of ash trees caused by the *Chalara fraxinea* fungus. The disease causes leaf loss and crown dieback in affected trees, and is usually fatal. Expected to spread throughout UK. **Further Information:** www.forestry.gov.uk/forestry/infd-8udm6s
Conifer root and butt rot *Heterobasidion annosum*	A decay fungus infecting coniferous trees. The fungus grows down through recently cut stumps, roots and across into any roots of nearby living trees. It can result in the death of pines on vulnerable sites, and decay in the lower stems of many other coniferous species. **Further Information:** www.forestry.gov.uk/fr/INFD-66RCKJ
Dothistroma (red band) needle blight	Primarily affects pine trees. Causes premature needle defoliation, resulting in loss of yield and, in severe cases, tree death. **Further Information:** www.forestry.gov.uk/fr/infd-6zckae
Dutch elm disease	Very serious disease caused by two related species of *Ophiostoma* fungi which are passed on by various elm bark beetles. You are very unlikely to have any mature elm trees in your wood as a result of this disease, though there are many elms in hedgerows that dieback after 15-20 years and then re-shoot. **Further Information:** www.forestry.gov.uk/fr/HCOU-4U4JCL

Disease	Notes
Oak decline / oak dieback	A complex disorder or syndrome in which several damaging agents interact to cause decline. These include oak leaf roller moth, oak mildew, and *phytophthora*. **Further Information:** www.forestry.gov.uk/fr/INFD-7B3BLF
Phytophthora austrocedrae on juniper	Recently found causing death of junipers in woodland in northern Britain. **Further Information:** www.forestry.gov.uk/pdf/phytophthora_austrocedrae_juniper_factsheet.pdf
Phytophthora on alder	Causes crown dieback and bleeding from stem tissue. **Further Information:** www.forestry.gov.uk/fr/INFD-737HUN
Phytophthora infected rhododendron	Rhododendron is usually an unwelcome invasive species and provides a host for the *Phytophthora* fungus that can affect adjacent trees. **Further Information:** www.forestry.gov.uk/fr/INFD-8F7BU3 and www.forestry.gov.uk/fr/INFD-73RAUP
Phytophthora ramorum (sudden oak death)	In spite of its name, much of the worst damage is to larch, but does affect other species including oak, beech, sweet chestnut, horse chestnut and Douglas fir. Causes stem cankers that frequently kill the host tree. Has killed thousands of hectares of larch. **Further Information:** www.forestry.gov.uk/pramorum

Invasive weed species

Thanks to our long history of overseas travel and horticulture, there are a large number of exotic invasive species in the UK. While very few of them will harm established trees, many will prevent seedlings from growing effectively by competing with them for light and nutrients. The main impact of invasive species is a loss of habitat diversity, as they will tend to take over the woodland floor and prevent other species from growing. If your woodland has previously been used for game shooting, you may find that two of the main culprits are already well established, rhododendron and snowberry having been commonly used in the past as cover for game birds.

While the term 'weed' is commonly used to mean a plant that is in an inconvenient place, there is a statutory definition as it is an offence to allow some species to spread from your land onto a neighbouring property. This is addressed by two acts of parliament:

- The Injurious Weeds Act (1959) which covers: Common ragwort, Spear thistle, Creeping or field thistle, Broad-leaved dock, and Curled dock
- The Wildlife and Countryside Act (1981) that includes more species such as: rhododendron, Himalayan balsam, giant hogweed and Japanese knotweed.

A quick note on ivy (which entirely concur with Julian's views): Ivy does not harm living trees, and shouldn't be considered to be a risk to healthy woodland. The only time trees are in danger from ivy growth is if the tree is *already* rotten and cannot take the additional weight, or if for some reason the ivy overshadows the tree's crown.

FURTHER INFORMATION

There is a useful summary of the law on invasive species at http://naturenet.net/law/invasive.html, and further information at www.gov.uk/weed-control-for-farmers, and at www.naturalengland.org.uk/ourwork/regulation/wildlife/enforcement/injuriousweeds.aspx.

Where to get more information

There are a number of different routes you can go down to find further information about the health of your woodland. In the first instance, if you're not sure, I'd suggest getting in touch with either your local Forestry Commission woodland officer or your local council tree officer. They will have good local knowledge about the factors that are likely to affect trees in your area and the sorts of species that may be affected. If you need further advice there are some other organisations that can help:

Forestry Commission plant health
The Forestry Commission has a plant health service responsible for enforcing legal requirements on biosecurity particularly in imports and exports. If you have any questions regarding plant health, then you can find their contact details here: www. forestry.gov.uk/website/forestry.nsf/byunique/infd–7nvert

Inspections of trees and woodland by Forestry Commission plant health inspectors play an important role in efforts to manage outbreaks of pests and diseases. There is an online document (www.forestry.gov.uk/pdf/Guidancetoplanthealthinspectors. pdf) giving guidance to plant health inspectors that should give you an idea of what to expect if your site it affected.

Forest Research Tree Health Diagnostic and Advisory Service
If you find that trees in your woodland are dying, and you suspect that they may be diseased, then the Forest Research Tree Health Diagnostic and Advisory Service is able to test samples. Further details of their service are available here www.forestry. gov.uk/fr/INFD–5UWEY6

A final thought

Most woodlands and most owners are unlikely to encounter major tree health problems. It's a bit like crime, fear of it is often far greater than the slight chance of being a victim. And also like crime, taking sensible precautions can mitigate the risk as well as help contain any problem that does arise.

11

Advice and Where to Get Help

In this last chapter to help you get started in your own wood, I again draw on my experience rather than simply present a dry list of information sources. These, such as addresses, other contacts and sources of information, are in the appendices. Help can come in many different ways. Even as a professional forester I have not been wanting for advice from others.

I've also summarised a few legal and taxation issues and comment on certification.

Neighbours

One of the best places to start is to ask a neighbour. To the south of our wood is a small organic market garden and Mike, the owner, willingly cuts my rides once a year, has cleared away rubbish dumped at our entrance, and generally keeps an eye on things. Next to Mike is Melvyn. He works coppice in traditional ways when not farming his smallholding, and there is probably nothing he doesn't know about local wildlife, not to mention other rural goings-on. Opposite Melvyn live the Armstrongs who readily phone if they've seen something amiss. Beyond them, Peter, who farms several fields, has kindly cut back our protruding hedge to allow combine harvesters easy passage along the lane so I get a free hedge trimming! Alan, whose 3 acre wood adjoins ours, is always passing on bits of local knowledge. Even Network Rail contribute to ride upkeep through the wood and cut back overhanging growth next to the railway. Neighbours are mostly a boon, and all of us like being asked for advice, for our opinion, and even for our help.

Reading matter

The ignorance of woodland lore is reflected by its absence in literature, though there is the joy of playing pooh-sticks, reading of the storm that blew down Owl's house, and other delights in A. A. Milne's *Winnie the Pooh* stories centred on One Hundred Acre Wood. One can, however, pick up a remarkable amount without really meaning to. Thomas Hardy's *Woodlanders* recounts much of coppicing, tree planting and the timber business in Victorian times, and who can fail to be impressed by Cobbett's *Rural Rides* or H. E. Bate's *Through the Woods* and his eye for detail in the cycle of woodland life as season follows season and altercation follows altercation with gamekeepers and the hunting fraternity. But, on the whole there is little written informally about woodlands and woodland work. My own efforts of telling the story of a wood in *A Wood of Our Own* and *What Happened to Our Wood* have, I believe, no antecedents as neither did Thomas Firbank's inspirational *I Bought A Mountain* relating his venture into sheep farming from scratch in the wilds of Snowdonia in the 1930s. That said, I would now (2015) want to add Ben Law's books, especially his story of his own Prickly Nut wood (see appendix on page 155).

More formally there are countless books about individual trees, how to identify them and what uses they have and increasingly they include chapters on tree management. Many gardening books contain sections about tree work, with some well-known authors like Alan Titchmarsh especially knowledgeable from their own experience. Alan owns a wood of 30 acres that is a delightful mix of new planting and mature woodland.

More formally still there are many books about forest and woodland management and silviculture, but few focus specifically on smaller woods apart from the late Ken Broad's almost encyclopedic *Caring from Small Woods* and Ben Law's informative personal perspective in *The Woodland Way*. A comprehensive guide into all aspects awaits an author though Chris Starr's *Woodland Management – a practical guide* gets nearest.

Periodicals and magazines that cover woodland work are either trade journals such as *Forest and Timber News* and *Forestry Journal*, or newsletters, magazines or journals associated with membership of societies, both professional and non-professional, of which there are many. Several are listed under societies and associations, but the most helpful for someone starting out are *Quarterly Journal of Forestry* and *Smallwoods* that are the journals of the Royal Forestry Society and Small Woods Association respectively.

When to get in touch with the authorities

Forestry Commission

This is the government department (or Natural Resources Wales – NRW) responsible for forestry matters. For the small woodland owner the Forestry Commission will be the most important contact as *all felling of trees, beyond very small quantities, must be covered by a current felling licence or an approved management plan.* Some kinds of *tree planting and regeneration are eligible for grant aid* as are some forms of woodland improvement work such as coppicing and provision for access or wildlife. Also the Forestry Commission or NRW may support wood fuel processing and supply.

Get in touch with your local office using the telephone directory or the Forestry Commission website www.forestry. gov.uk. You will find them almost a 'one-stop-shop' for forestry and woodland matters, for the few minutes they can spare you, and the local woodland officer may occasionally be able to fit in a brief visit to your wood to discuss plans and ideas.

Local authorities

Many local authorities employ tree or woodland officers who may be able to give advice. Sometimes a district or county may offer grants to help with certain woodland operations. For many years Hampshire encouraged the restitution of neglected hazel coppice as part of their countryside policy. Their forestry officer visited our wood to look at the neglected hazel, but wasn't impressed with its stocking and it didn't merit grant aid! Tree and woodland officers are usually found in the Planning Department or with Parks and Gardens.

Local authorities are also responsible for Tree Preservation Orders and should always be contacted in connection with such matters. At the moment woodland operations do not come under planning control, though building and all related works do.

Other bodies

The Department for Environment and Rural Affairs (DEFRA) and their Scottish and Welsh equivalents may grant aid for farm woodland planting, often in conjunction with the Forestry Commission. The new Common Agricultural Policy from 2016 provides capital grants for woodland management plans and some tree health issues and schemes for supporting farm and forestry businesses.

Natural England, Scottish Natural Heritage and Natural Resources Wales may support some woodland work directed towards conservation improvement and, of course, will be directly involved if a woodland is a Site of Special Scientific Interest (SSSI) or is part of a nature reserve.

National Park Authorities will have an interest in woodlands in their area.

Joining societies and associations

There are numerous societies and voluntary bodies concerned with trees, woodlands and forests. I highlight here the ones of most interest to the small woodland owner and would encourage joining one or more. Addresses will be found in the Appendix.

Except for two bodies noted later, all forestry associations and societies are open to members of the public without pre-conditions. *The Royal Forestry Society of England, Wales and Northern Ireland* (who have helped sponsor this book) and *The Royal Scottish Forestry Society* hold regular field meetings by region that are a delightful way of acquiring forestry knowledge and meeting like-minded people. They issue quarterly periodicals. This is also true of *Confor* who represent the interests of large and small private woodland owners who particularly have an interest in commercial timber production. Both *The Small Woods Association* (SWA) and the *Small Woodlands Owners Group* (SWOG) are excellent groups to join since they link together many different parties, individuals and organisations, run workshops and training sessions, all with a special interest in smaller woodlands.

Other groups who organise visits to woodlands and have regional meetings are *The International Tree Foundation* and *Woodland Heritage*. Membership of *The Woodland Trust, The*

National Trust and any one of the array of conservation bodies, local and national, like *The British Trust for Conservation Volunteers,* will all add further opportunities to learn. *The Tree Council* is active in promoting tree planting and tree care – they instigated the annual tree week in early December – and also have a network of volunteer tree wardens who take a special interest in tree matters in their locality. They publish an attractive 'glossy' called *Tree News.*

The Institute of Chartered Foresters is the main professional body and one is only eligible to join through a professional membership entry process. Most consultants are members and are 'chartered foresters'. *The Royal Institution of Chartered Surveyors,* similarly only open to membership by examination, has a forestry group who are also consultants.

Calling in the consultant

I've stressed more than once that even an hour's visit by a qualified forestry consultant can revolutionise your thinking about a wood. Its possibilities and potential will become apparent, valuable timber trees identified, and management options discussed. For any wood of more than a few trees at least one visit at some stage by a knowledgeable person with a forestry qualification will repay the outlay.

Calling on a professional is especially important if there is concern over safety, say of a tree next to a public right-of-way. The local authority tree officer will also be a good person to contact.

A list of forestry consultants (chartered foresters) is available on the *Institute of Chartered Foresters* website www.charteredforesters.org.

Using contractors

All of us have probably had a bad experience with contractors with a job poorly done or a feeling we have been 'ripped off'. This can happen in forestry work, but no more so than other industries. The best way to guard against this is to follow someone else's recommendation, or refer to the *Confor* who publish a list of members who undertake contracting.

As you will know by now, in my own wood I have sold all significant quantities of timber by tender and the buyer has employed felling contractors to cut and extract the timber. I have had few problems and, indeed when felling large oaks in part of the wood called Nain's Copse, the standard of work by the contractors was nothing short of exceptional. No neighbouring tree was barked, the logs were left beautifully presented, and the job was done cleanly and safely in just a few days.

In general consider employing a contractor for all significant tree felling (and tree pruning if safety is a concern) or if you are planting and doing maintenance work involving more than, say 100 trees, fencing work, or road and track construction. Of course, this entirely depends on your circumstances, your skills, and time available! Remember to make sure your contractors are properly covered for insurance, are trained for the equipment they use, and comply with current health and safety legislation. As mentioned earlier, a woodland owner may be deemed the 'Forest Works Manager' under HSE guidelines and so carries more responsibility than a domestic homeowner engaging a contractor to work on the house.

Taxation and related issues

Since 1988 forests and woodland have been largely taken out of the taxation environment. No income tax is payable on timber and wood products sold from a woodland, apart from the commercial growing of Christmas trees. So, if you can sell a fine oak for hundreds of pounds, no tax has to be paid. Indeed, the Inland Revenue (HMRC) simply won't be interested. Grants for woodland work are also tax-free.

It gets better. The value of standing trees and woods is not liable for capital gains tax and never has been.

It gets even better. Woodland will not normally attract inheritance tax either. When a woodland investment has been held for more than two years it qualifies for 100 per cent business relief from inheritance tax. And even if the wood is made over as a gift, after the two year qualifying period the business relief is passed to the new owner. The usual rule that a donor must survive seven years after making an absolute gift for it to escape inheritance tax does not apply.

This favourable financial treatment is a way that the government helps owners of forests and woodlands. It realistically reflects the generally poor return from growing timber and the long timescales and commitment required while recognising that in the wider landscape, trees and woodlands add immeasurably to beauty, amenity and wildlife. You will never get rich owning and managing forests: the government agrees, freeing you of most taxes.

The above provisions apply to <u>commercially managed</u> woodlands i.e. you have sold some timber or manage the wood with the intention of producing some for sale at some point in the future. It is, after all, a business relief. That said, most woods, even small ones, are managed for many purposes. Provided you make some sales from your wood, and keep a note of them, or have this as a management intention in five or even 10 years time, then you will probably satisfy the criteria of 'commercially managed'. If the taxation side of things is an important aspect of your buying or owning a wood, do check for yourself.

Certification

In the last 20 years we have seen the rapid rise of independent bodies to certify that good standards of woodland management are carried out. The aim is to ensure sustainable management to perpetuate forests and woodlands. It began in the tropics as one way to help reduce forest destruction and deforestation, but now embraces many countries. Most of us will have seen the FSC (Forestry Stewardship Council) or PEFC (Programme for Endorsement of Forest Certification) logos on timber bought from DIY stores.[1]

How does this affect the small woodland owner? It may only be an issue if you want to sell timber. Increasingly buyers require assurance that the wood comes from a certified forest or woodland. You probably won't get a better price if your wood is certified, but a more ready sale.

For the small owner the difficulty with certification is

---

1 Both of these organisations try to address the issue of ensuring that their certified timber comes only from sustainable sources. Though imperfect, they give some measure of assurance. In Britain we have in addition the UK Woodland Assurance Scheme (UKWAS) which sets minimum standards of management and must be complied with to be eligible for Forestry Commission grants.

that it is expensive. The certifying organisation charges for the evaluation they do, and then for their annual or biannual checking to remain certified. What is required is for your wood and its management to comply with the UK Forestry Standard.

Because of the disproportionately heavy cost on a small owner, many such owners, including me, have not yet taken the step of certification. I think I will soon and will join a group scheme for small owners to club together, since conditions are a little less stringent to comply with than those obtaining for large commercially-run forests. The Soil Association is one body exploring such an approach to help the small owner. Organisations like 'Myforest' can offer plenty of help.

I don't want to finish this book on a negative note because the grand aim of certification is to ensure better run forests and woodlands, an aim with which most will surely agree. We want to see woodlands sustainably managed, and I hope this book has helped you get started in yours. At the end of the day we want to be able to say that we have been good stewards of what, for me as a Christian, God has entrusted to our care – one of the lovely woods that so bless Britain's countryside.

John White's lovely sketch of beeches – good luck with your wood.

Glossary

Some of the terms used in this book may be unfamiliar. A list can never be complete, but I hope the simplified definitions below are a help.

acre unit of area equal to 0.4 hectares

agroforestry growing trees and food (plants or animals) together

ancient woodland land which has been wooded continuously since before 1600

biodiversity (biological diversity), variety and abundance of species

biomass crops grown for fuel, commonly as short rotation coppice

brashing removal of lower branches by pruning to allow access

breast height 1.3m above ground and position where stem diameter is measured

butt bottom most part of a tree trunk and usually largest log

canopy branches and leaves of trees that spread overhead and shade woodland floor

cleaning removing woody growth, climbers and other unwanted plants from a developing stand of young trees

clearfell felling of all trees in a stand in one go

continuous cover forestry managing forest so that there is always some tree cover

coppice using shoots that emerge from cut stumps to regenerate woodland

crown the aerial assemblage of branches and leaves of a tree

flush burst of new growth in the spring as buds expand and open

habitat the place where animals or plants live

hectare unit of area equal to 2.47 acres

native occurs naturally and not as a result of introduction by man

natural regeneration regenerating forest from seed fall or coppice i.e. not by planting

niche the home (environment) suited to particular organisms

pollard a tree regularly cut 2-4m above ground for small sized branch wood that is out of reach of browsing by deer and livestock

ride track through a wood usually wide enough for vehicles

rotation period of time between planting and felling, or age when trees are felled

roundwood wood used 'in the round' – not sawn – such as stakes or sold like this e.g. firewood

sawlog a big enough log suitable for sawing into planks

short rotation coppice coppicing on a cycle (rotation) of less than 10 years, often 2-4 years

silviculture c.f. agriculture, the husbandry, care and tending of woods and forests

snag American word for dead trunk left standing for wildlife benefit

stand a community of trees with similar characteristics and managed as one unit

standing sale sale of trees where they are growing and before they are felled

swipe operation of cutting weeds and grass in a ride

thinning from time to time, removing some trees from a stand to favour others

underwood woody undergrowth beneath a tree crop, may sometimes be managed. Coppice beneath standards may sometimes be called underwood

wildwood pre-existing woodland before the influence of humans

Acronyms

We are beset by shorthand use of acronyms, here are some used in the book or which you might come across.

AGLV Area of Great Landscape Value

AONB Area of Outstanding Natural Beauty

ASNW Ancient Semi-Natural Woodland

BCTGA British Christmas Tree Growers Association

BDS British Deer Society

BES British Ecological Society

BTCV British Trust for Conservation Volunteers

CLA Country Land and Business Association

Confor (name of what was once Timber Growers Association)

CPRE Council for Protection of Rural England

CROW Countryside and Rights of Way act

DEFRA Department of Environment, Food and Rural Affairs

EWGS England Woodland Grant Scheme (check for changes under Rural Development Programme – **RDPE**)

FC Forestry Commission

FE Forest Enterprise

FSC Forest Stewardship Council

FTA Forestry and Timber Association

FWAG Farming and Wildlife Advisory Group

HAP Habitat Action Plan

HSE Health and Safety Executive

ICF Institute of Chartered Foresters

NE Natural England

NRW Natural Resources Wales (includes former Forestry Commission Wales)

PAWS Plantations on Ancient Woodland Sites

PEFC Programme for Endorsement of Forest Certification

RFS The Royal Forestry Society of England, Wales and Northern Ireland

RSFS The Royal Scottish Forestry Society

SEERAD Scottish Executive Environment Rural Affairs
　　Department
SFGS Scottish Forestry Grant Scheme
SHAI Site of High Archaeological Importance
SNCI Site of Nature Conservation Interest
SNH Scottish Natural Heritage
SRC Short Rotation Coppice
SSSI Site of Special Scientific Interest
SWA Small Woods Association
SWOG Small Woodland Owners Group
TPO Tree Preservation Order
UKFS United Kingdom Forestry Standard
WDA Welsh Development Agency
WGS Woodland Grant Scheme

Notes of Common Woodland Trees
and a Few Woody Shrubs and Climbers

[Refer to pages 127–130 for pest and disease threats to individual tree species.]

Common name	Scientific name	Silvicultural notes	Soil[1] and Site needs

Native Broadleaves

Common name	Scientific name	Silvicultural notes	Soil[1] and Site needs
Alder	*Alnus glutinosa*	Hardy tree, tolerates flooding	Wet soils, stream and river sides
Ash	*Fraxinus excelsior*	Frost tender, stems often fork. Light demander	Rich, moist soils, OK on chalk soils. Not exposed sites
Beech	*Fagus sylvatica*	Tolerates shade. Prone to squirrel damage	Well drained loams. Avoid heavy soils
Birch	*Betula pubescens* *Betula pendula*	Arises naturally wherever mineral soil is exposed	Acid, sandy and gravelly soils. Grows in uplands.
Elm	*Ulmus* spp.	Trees up to 10m still common in hedgerows	Fertile, deep soils
Field maple	*Acer campestre*	Wood margins in South of Britain	Deep heavy soils, copes with chalk
Hornbeam	*Carpinus betulus*	Very hardy. Needs high stump to coppice well	Heavy clay soils
Holly	*Ilex aquafolium*	Minor component of many woods	Where beech and oak grow well
Lime	*Tilia platyphyllos* *Tilia cordata*	Component of native woods, good in hedgerow	Fertile soils
Oak (Common) (Sessile)	*Quercus robur* *Quercus petraea*	Commonest native broadleaf. Needs full light to grow.	Tolerant of many soils, grows well on clays

Height and growth[2]	Uses	Other notes
20m, moderate growth rate	Land restoration. Amenity. Turnery	Enriches soil by fixing nitrogen
30m, moderately fast growing when young	High quality hard-wood. Good firewood	Sets seeds most years, natural regen. common, ash dieback threat
30m+, slow to moderate growth, mature at 100+ y	Difficult to grow furniture quality. Charcoal.	Often found on chalk and limestone – suffers chlorosis
20m+, fast when young, rarely lives longer than 80 y	Amenity tree in landscape. White wood for turnery	Two closely related species.
30 m, but few big trees owing to DED	Landscape. Coffins, Windsor chairs	Dutch elm disease (DED) still killing trees aged more than 20y
15m+, slow to moderate growth	Good for hedges. Conservation	Winged seeds a lovely port red
20m, moderate growth	Dense wood, good for charcoal	Most common in Southeast England
15m, slow growing	Wood is good for carving & turning	Usually self-seeding
30m+, moderate growth	Conservation planting. Turnery and carving	Two native species, hybrid common in avenues
30m, slow growth, mature at 120y+, long lived	Sound timber is in demand. Conservation	Species similar. Both suffer squirrel damage

Common name	Scientific name	Silvicultural notes	Soil[1] and Site needs
Poplar (Black) (Aspen)	*Populus nigra* *Populus tremula*	Susceptible to canker and rust	Deep, fertile soils. Not exposed sites
Rowan	*Sorbus aucuparia*	Very hardy	Upland sites
Whitebeam	*Sorbus aria*	Early coloniser in scrub on chalk	Lowland sites, inc. chalky soils
Wild cherry or gean	*Prunus avium*	Use in mixture with other species	Mostly lowlands inc. chalky soils
Wild Service Tree	*Sorbus torminalis*	Minor component of native woods	Lowland sites, at home on clays
Willow	*Salix* spp.	Sallow (*S. caprea*) very widespread as early coloniser	Varied

Woody Shrubs and Climbers

Alder buckthorn	*Frangula alnus*	Widespread	Wet, acid soils
Bird cherry	*Prunus padus*	Often defoliated by ermine moth	Upland woods and streams in North
Blackthorn	*Prunus spinosa*	Suckers heavily	
Clematis or old man's beard	*Clematis vitalba*	Woody climber, can overwhelm	Sure indicator of chalky soils
Purging buckthorn	*Rhamnus carthatica*	Most common in south and east	Thrives on poor chalky soils
Dogwood	*Cornus sanguinea*	Mainly South England	Rich, chalky soils
Elder	*Sambucus nigra*	Throughout UK	Nitrogen-rich soils
Hawthorn	*Crataegus monogyna*	Widespread	All but the poorest soils
Hazel	*Corylus avellana*	As underwood or pure coppice	Heavier soils, acid to chalky

Height and growth[2]	Uses	Other notes
20m+, often fast growth	Little commercial use. Conservation	Black poplar now rare
15m, moderate	Amenity	Birds eat berries
20m, moderate	Amenity	Light demanding
25m, fast when young	Fine cabinet wood. Amenity	Lovely blossom in spring. Suckers
20m, slow to moderate growth	High conservation value	Indicates ancient woodland site
Many are very fast	Not timber, except cricket bat willow *S. alba* 'Coerulea'	Five native species. Purple Emperor feeds on sallow

	Excellent charcoal	Has no thorns
Small tree	Conservation	Birds eat berries
	Cover for wildlife	Impenetrable!
	A nuisance!	Why 'Travellers Joy?'
Small tree	Hard reddish-brown wood	Berries have a powerful affect!
	Very hard wood	Birds carry seeds
	Elderflower wine	Hollow stems
Small tree	Excellent hedge	The wonderful 'may' flower
8m, fast initial growth from stool	Thatching spars, hurdles. Cover	Many neglected coppices.

Common name	Scientific name	Silvicultural notes	Soil[1] and Site needs
Honeysuckle	*Lonicera periclymenum*	Woody climber in glades and rides	Most soils
Ivy	*Hedera helix*	Woody climber	Throughout UK
Privet	*Ligustrum vulgare*	Mainly South England	Chalky soils
Spindle tree	*Euyonomus europeaus*	Mainly South England	Commonest on chalky soils
Wayfaring tree	*Viburnum lantana*	Mainly South England	Dry chalky soils

Introduced Broadleaves

Poplars	*Populus* spp.	Disease resistant cultivars only. Put trees far apart	Fertile soils, sheltered sites
Sweet chestnut	*Castanea sativa*	Light demanding, flowers heavily, nuts disappointing	Acid, well drained soils. Warm, sunny conditions
Sycamore	*Acer pseudoplatanus*	Hardy tree, occurs throughout UK. Regenerates freely	Tolerant of wide range of soils. Stands exposure
Walnut	*Juglans regia*	Frost tender, usually grown as single trees	Rich well-drained soil. Warm, sheltered sites

Native Conifers

Juniper	*Juniperus communis*	Widespread but declining in extent	Tolerant of a range of soils
Scots pine	*Pinus sylvestris*	Regenerates easily on heathland	Any well drained acid soil
Yew	*Taxus baccata*	Pure stands rare, tolerates shade	Well drained soils, inc. chalky

Height and growth[2]	Uses	Other notes
	Ash/hazel walking sticks – twisted on	Lovely fragrance on still evenings
	Winter cover, late flowers help bees	Need not remove
	Hedging and cover	Evergreen
Rarely a tree	Fine turnery – pegs, needles	In November, bright pink berries
Not really a tree		Beside droveways
30m+, very fast	Biomass, short rotation coppice. Light timber	Some agroforestry potential
30m, moderate, fast when young coppice	Durable wood for palings and posts. Oak-like timber	Extensive coppice in Southeast England. Roman introduced
30m, moderate growth rate	General purpose hardwood. Some high value butts	Suffers squirrel damage. Bark good for lichens
20m, slow	Cabinet wood	Rarely produces edible fruit. Roman introduced
15m, slow	Conservation, esp. in uplands	Food source and cover for birds
30m, moderate growth rate	Timber. Conservation (Scotland)	Severe chlorosis on chalky soils
20m, slow, very long lived	Amenity, historic interest	Bark, berries and foliage poisonous

Introduced Conifers

Common name	Scientific name	Silvicultural notes	Soil[1] and Site needs
Corsican pine	*Pinus nigra* var. *maritima*	Light demanding. Suffers red-band needle blight	Well-drained soils on lowland sites
Douglas fir	*Pseudotsuga menziesii*	Tolerates some shade, continuous cover forestry	Fertile well-drained soils. Sheltered sites
European larch	*Larix decidua*	Light demanding, prone to canker and ramorum	Well drained soils
Firs	*Abies* spp.		
Japanese larch	*Larix kaempferi*	Light demanding Devastated by ramorum	Tolerant of wide range of soils
Lodgepole pine	*Pinus contorta*	Hardy, for uplands	Wet infertile soils
Norway spruce	*Picea abies*	Moderately hardy	Heavy soils, drier sites in East
Sitka spruce	*Picea sitchensis*	Most widely planted conifer, at risk of windthrow	All soils incl. peats and gleys if rain more than 1,000 mm
Western hemlock	*Tsuga heterophylla*	Shade tolerant. Regenerates freely	Well drained acid soils in lowlands
Western red cedar	*Thuja plicata*	Best in mixture	Well drained soils, tolerates chalk

1 Soil – reference to chalk or chalky soil means any calcareous soil with lime or chalk fragments or that is very shallow over chalk or limestone and is alkaline of pH more than 7

2 Height and growth – very approximate guide to final heights typical in UK; fast growth – up to 1m in a year, moderate 40-70cm, slow less than 40cm

Height and growth[2]	Uses	Other notes
35m, moderate to fast, mature in 40+ years	Construction timber	Some tolerance of chalk/lime in soil
40m+, fast, mature at 40y+, very long lived	Fine construction timber	Readily browsed by deer when young
30m, moderate growth rate	Stakes and rustic. Sheds and fences	Forms hybrid with Japanese larch
		Minor importance
30m, fast when young	Stakes and rustic. Sheds and fences	Stems corkscrew, use hybrid
25m, moderate	Industrial uses	Little now planted
30m, moderate to fast	Timber, Christmas trees	Suffers from great spruce bark beetle
35m, fast, mature at 40y+, long lived	Good pulpwood, and construction timber	Never use as Christmas trees, needles prickly
30m+, fast	Poorer timber than other conifers	Regeneration can be a nuisance
30m, moderate to fast, long-lived	Very durable timber	Cigar-shaped trees can look ugly

Further Reading

Individual trees and forests

Harris, E, Harris, J and James, NDG (2003) *Oak: A British History*, Windgather Press.

Hinde, T (1985) *Forests of Britain*, Victor Gollancz.

Hyde, HA *Welsh Timber Trees – Native and Introduced*, National Museum of Wales.

Marren, P (1992) *The Wild Woods: A regional Guide to Britain's Ancient Woodlands*, David and Charles.

Milner, E (2011) *Trees of Britain and Ireland*, Natural History Museum, London.

Mitchell, AF (1992) *A Field Guide to Trees of Britain and Northern Europe*, Collins.

More, D and Johnston, O (2004) *Collins Tree Guide*, Collins.

Thomas, P (2000) *Trees: Their Natural History*, Cambridge University Press.

White JEJ (1995) *Forest and Woodland Trees of Britain*, Oxford. (By our illustrator.)

Woodland and forestry operations

Agate, E (ed). (2002) *Woodlands: A Practical Handbook*, BTCV.

Broad, K (1998) *Caring for Small Woods*, Earthscan, London.

Blyth, J, Evans, J, Mutch, WES and Sidwell, C (1991) *Farm Woodland Management*, 2nd ed., Farming Press, Ipswich.

Claridge, J (ed). (2003) *So, You Own A Woodland – Getting to Know Your Wood and Looking After it*, Forestry Commission England.

Evans, J (1984) *Silviculture of Broadleaved Woodland*, Forestry Commission Bull. 62, HMSO.

Harmer, R, Kerr, G and Thompson, R (2010) *Managing Native Broadleaved Woodland*, Forestry Commission, TSO.

Helliwell, R (2006) *Fundamental Woodland Management*, Small Woods Association.

Hibberd, BG *Forestry Practice*, Forestry Commission Handbook 6, Stationery Office.

Kerr, G and Evans, J (1993) *Growing Broadleaves for Timber*, Handbook 6, Forestry Commission, HMSO.

Matthews, R and Mackie, E (2006) *Forest Mensuration – A Handbook for Practitioners*, Forestry Commission, TSO.

Mutch, WES (2008) *Tall Trees and Small Woods – How to Grow and Tend Them*, Mainstream Publishing, Edinburgh and London.
Small Woods Association (2006) *A Marketing Guide for Owners of Small Woods*, Small Woods Association and Forestry Commission, England. (Free.)
Starr, C (2013) *Woodland Management – A Practical Guide*, 2nd ed., The Crowood Press.
Stokes, J and Hand, K (2001) *The Good Seed Guide*, The Tree Council, London.

Natural history and conservation in woodlands

Blakesley, D and Buckley, GP (2010) *Managing Your Woodland for Wildlife*, Pisces Publications, Newbury.
Harris, E and Harris, J (2003) *Wildlife Conservation in Managed Woodlands and Forests*, Research Studies Press.
Peterken, G F (1996) *Natural Woodland*, Cambridge University Press.
Rackham, O (2003) *Ancient Woodland: Its History, Vegetation and Uses in England*, Castlepoint Press.
Rackam, O (2001) *Trees and Woodland in the British landscape*, Weidenfeld and Nicholson.
Rackham, O (2012) *Woodlands*, Collins (New Naturalist Library).
Schama, S (2004) *Landscape and Memory*, Harper Perennial.

Using wood as fuel

Laughton, C (2006) *Home Heating with Wood*, CAT Publications.
Rolls, W (2013) *The Log Book: Getting the Best From Your Woodburning Stove*, Permanent Publications, East Meon.

Accounts of individual woodlands but with much else of relevance

Evans, J (1995) *A Wood of Our Own*, Oxford, reprinted 2003, Permanent Publications.
Evans, J (2002) *What Happened to Our Wood*, Patula Books, Basingstoke.
Law, B (2001) *The Woodland Way – A Permaculture Approach to Sustainable Woodland Management*, Permanent Publications, East Meon.
Law, B (2013) *Woodsman: Living in a Wood in the 21st Century*, Harper Collins.

Organisations to Join
of Relevance to Owners
of Small Woods

British Trust for Conservation Volunteers (BTCV)
One of the best bodies to join for gaining practical hands-on experience.

Confor
Forestry's equivalent of the NFU with focus on owners of larger woodlands, still much of relevance to the smaller owner. Have regional groups that hold useful field visits. Publishes *Forestry and Timber News.*
Confor, 59, George Street, Edinburgh EH2 2JG
(0131 240 1410) www.confor.org.uk

Local Wildlife Trusts
Information, surveys and management plans. Site visits and local practical work.
Local Wildlife Trust (01636 670001) www.wildlifetrusts.org

Royal Forestry Society of England and Wales (RFS)
Largest society producing the very readable *Quarterly Journal of Forestry*, excellent website and monthly e-news digest, holds regular site meetings, good regional structure throughout England and Wales. After Small Woods Association join RFS, you'll benefit greatly. It is not expensive.
Royal Forestry Society of England and Wales (RFS), The Hay Barns, Home Farm Drive, Upton Estate, Banbury OX15 6HU (01295 678588) www.rfs.org.uk

Royal Scottish Forestry Society (RSFS)
Scottish equivalent of RFS.
Royal Scottish Forestry Society, St Leonards, Maxwell Lane, Kelso, TD5 7BB (01387 383845) www.rsfs.org

Small Woodlands Owners Group (SWOG)
Formed and run by owners of small woods through support from Woodlands.co.uk. Excellent networking, field visits

and training, plus regular newsletter and a useful source of information about courses recommended by members. Great for sharing experience and learning from others. Join either or both SWOG and SWA (below).
www.swog.org.uk

Small Woods Association (SWA)
Probably the 'must join' organisation with a regular magazine (*Smallwoods*), relevant publications (some free), site meetings and training days for everyone interested in small woodland management.
Small Woods Association, The Green Wood Centre, Station Road, Coalbrookdale, Shropshire TF8 7DR
(01952 432769) www.smallwoods.org.uk

The Tree Council
Many organisations belong to the Tree Council. Concerned with management and conservation of trees and woodlands. Promote annual tree week and 'Walk in the woods'. Publishes a lovely glossy: *Tree News*

The Woodland Trust
A bit like the National Trust except focusing on woodlands, both existing and creating new ones. Free access to their woodlands in UK – look around and get ideas. Some opportunities for volunteer work.
Woodland Trust, Kempton Way, Grantham, Lincs. NG31 6LL
(01467 581111) www.woodland-trust.org.uk

Woodland Heritage
Emphasises managing woodlands for their timber so that the best home-grown timber is available and used in UK. Annual magazine, *Woodland Heritage*, excellent field meeting.
Woodland Heritage, PO Box 168, Haslemere GU26 1XQ
www.woodland.heritage.org.uk

Some Useful Addresses

Coppice/coppicing websites
www.coppice.co.uk
www.coppicegroup.wordpress.com

Forestry Commission HQ
Silvan House, 231, Costorphine Rd. Edinburgh. EH12 7AT
(0131 334 0303) www.forestry.gov.uk

Forestry Contracting Association
Tigh na Creag, Invershin, Lairg, Sutherland IV27 4ET
(0870 042 7999) www.fcauk.com

Institute of Chartered Foresters (ICF)
59 George Street, Edinburgh EH2 2JG
(0131 240 1425) www.charteredforesters.org

Natural England
Foundry House, 3 Millsands, Riverside Exchange, Sheffield S3 8NH
(0854 600 3078) www.naturalengland.gov.uk

Natural Resources Wales (NRW)
Ty Cambria, 29 Newport Road, Cardiff CF25 0TP
(0300 065 3000) www.naturalresourceswales.gov.uk

Scottish Natural Heritage (SNH)
Great Glen House, Leachkin Road, Inverness IV3 8NW
(01463 725000) www.snh.gov.uk

The Conservation Volunteers (TCV)
Sedum House, Mallard Way, Doncaster DN4 8DB
(01302 388 883) www.tcv.org.uk

Tree Council
71 Newcomen Street, London SE1 1YT
(0207 407 9992) www.treecouncil.org.uk

Woodlands.co.uk
19 Half Moon Lane, London SE4 9JU
(0207 737 0070) www.woodlands.co.uk

Index

[Note: additional named organisations, and tree and shrub species will be found in the relevant appendix and are not indexed here.]

 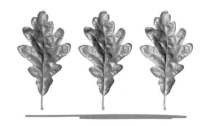

RFS
ROYAL FORESTRY SOCIETY

Welcoming you to woodland management

The RFS is the largest and longest established educational charity dedicated to promoting the wise management of woodlands and trees in England, Wales and Northern Ireland.

We offer:

- **a unique mix of opportunities**
  to meet and learn from others passionate about woodland management and to be part of the wider conversation around the future of our woodlands and forests.

- **a warm welcome**
  whether you have just bought your first woodland or have been managing one for some time, our members will share their own experiences on nearby sites.

Members can attend up to 100 woodland meetings across England, Wales and Northern Ireland on a huge variety of topics.

RFS members also enjoy a whole range of other benefits and the opportunity to support our educational work to advance the long term health of the country's woods and trees.

"I have learnt more attending RFS field meetings than from any other forestry source" RFS member

Find out more and join us:

online at **www.rfs.org.uk**

call us on **01295 678624**

or email

membership@rfs.org.uk

Enjoyed this book? You may also like these from Permanent Publications

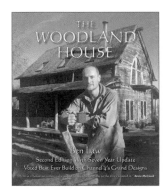

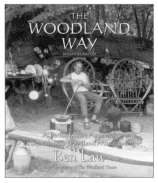

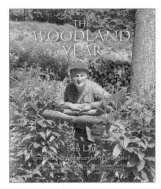

The Woodland House	**The Woodland Way**	**The Woodland Year**
Ben Law	Ben Law	Ben Law
£16.95	£24.95	£24.95
As seen on Channel 4's Grand Designs, this is a visual guide to how Ben built his outstandingly beautiful home in his own woods.	This radical book presents an immensely practical alternative to conventional woodland management using a permacultural approach.	Packed with stunning colour photographs, this book is an intimate month-by-month journey through Ben's yearly cycle of work.

Our titles cover: permaculture, home & garden, green building, food & drink, sustainable technology, woodlands, community, wellbeing and so much more

Available from all good bookshops and online retailers, including the publisher's online shop:
https://shop.permaculture.co.uk
with 10% off the RRP on all books

Our books are also available via our American distributor, Chelsea Green:
www.chelseagreen.com/publisher/permanent-publications

Permanent Publications also publishes *Permaculture Magazine*